AF598586

This book aims to familiarise the reader with the essential properties of the chaotic dynamics of Hamiltonian systems. It includes unique material on separatrix chaos, small nonlinearity chaos, fractional kinetics, and discussions on Maxwell's Demon and the foundation of statistical physics. Special mathematical tools not typical of physics are avoided.

The book is ideally suited for all those who are actively working on the problems of dynamical chaos. It introduces the physicist to the world of Hamiltonian chaos and the mathematician to actual physical problems. The material can also be used by graduate students.

*George M. Zaslavsky is Professor of Physics and Mathematics at New York University. He is known for his work on classical and quantum chaos and has published several books in the field.*

# PHYSICS OF CHAOS IN HAMILTONIAN SYSTEMS

**George M. Zaslavsky**

*Department of Physics, New York University*
*Courant Institute of Mathematical Sciences*

ICP

Imperial College Press

*Published by*

Imperial College Press
203 Electrical Engineering Building
Imperial College
London SW7 2BT

*Distributed by*

World Scientific Publishing Co. Pte. Ltd.
P O Box 128, Farrer Road, Singapore 912805
*USA office:* Suite 1B, 1060 Main Street, River Edge, NJ 07661
*UK office:* 57 Shelton Street, Covent Garden, London WC2H 9HE

**Library of Congress Cataloging-in-Publication Data**
Zaslavskiĭ, G. M. (Georgiĭ Moiseevich)
Physics of chaos in Hamiltonian systems / George M. Zaslavsky.
p. cm.
Includes bibliographical references and index.
ISBN 1-86094-052-8 (alk. paper)
1. Hamiltonian systems. 2. Chaotic behavior in systems.
3. Mathematical physics. I. Title.
QC20.7.H35Z38 1998
514'.74--dc21 98-23901
CIP

**British Library Cataloguing-in-Publication Data**
A catalogue record for this book is available from the British Library.



Printed in Singapore by Uto-Print

# Preface

The term *chaos* is often used to describe the phenomena in which the system's trajectories are sensitive to the slightest changes in initial conditions. In reality, the properties of such motion resemble those of random motion. If one restricts oneself to Hamiltonian systems with area-preserving dynamics, the above definition of chaos would appear perplexing. Can a trajectory be chaotic at times and "regular" for the rest of the time? How does a regular trajectory transform itself into a chaotic one? And what type of randomness characterises chaotic dynamics? Fortunately, clear definitions exist for regular and chaotic motions. Conditionally periodic motion is an example of regular dynamics. The examples of "ideal" chaotic motion refer to the motion on the negative curvature surface and the so-called Anosov systems. However, real physical systems or their simplified models are very different from the ideal models on chaos. A good example is that of a pendulum disturbed by a periodic (non-random) force.

Our understanding of chaos is fraught with the following difficulties. The motion known as chaotic occupies a certain area (called stochastic sea) in the phase space. In ideal chaos, the stochastic sea is occupied in a uniform manner. This is, however, not the case in real systems or models. The phase space contains many "islands" which a chaotic trajectory cannot penetrate. Initially, it had appeared that the effects of islands could be easily accounted for by simply changing the phase volume of the stochastic sea. We now believe that this is not true and the important properties of chaotic dynamics are in fact determined by the properties of the motion near the boundary of islands.

The difficulties in understanding Hamiltonian chaos can also be described in an informal way. While regular and chaotic motions possess some degree of uniformity (monotonity), which is used in their definitions, real chaotic motion boasts intermediate properties (between regular and chaotic motions) that have not been accurately defined and formulated. Therefore, we can neither use the KAM theory (as the conditions of non-degeneracy are violated in most cases) nor the Sinai's method of Markov partitions or related techniques (because the correlations do not decay exponentially and the Markov property is violated in some areas) and the estimates of Arnold diffusion (since a much faster diffusion takes place).

This book considers many of the difficulties described above in analysing Hamiltonian chaos in real systems. The reader is treated to the unconventional application of the fractal dimension to space-time objects, different versions of the renormalisation group method, fractional kinetics, and Poincaré recurrences theory as well as the more traditional applications of the Poincaré and separatrix maps.

This book is useful to the reader who has an undergraduate degree in physics. It does not include any methods that are beyond the standard mathematical physics techniques except for some fractional calculus which is provided in a special appendix. It is useful to physicists, engineers and those who are interested in the current problems in chaos theory and its applications. While mathematicians will not be able to find any rigorously proven results here, they will learn about the challenges of "real physics". Towards that end, we have included many examples of numerical simulations. Some of the examples were extracted (and updated) from the author's previous works. The material covered is based partly on the course offered by the author at the Courant Institute of Mathematical Sciences and the Department of Physics in New York University. Much of it, however, is new and can be used as a source of information on the new and emerging directions in modern chaos theory applied to physical problems.

I would like to thank those who have helped to make this book possible: Sadrilla Abdullaev, Valentine Afraimovich, Sadruddin Benkadda, Mark Edelman, Serge Kassibrakis, Jossy Klafter, Leonid Kuznetsov,

Boris Niyazov, Alexander Saichev, Don Stevens, Hank Strauss, Michael Shlesinger, Harold Weitzner and Roscoe White, all of whom are co-authors of common publications; Valentine Afraimovich, Leonid Bunimovich, John Lowenstein, Victor Melnikov, Anatoly Neishtadt, Yasha Pesin, Vered Rom-Kedar, Michael Shlesinger and Dimitry Treschev for their numerous discussions; Mark Edelman for his help in the preparation of the figures; Herman Todorov and Leonid Kuznetsov for their help in editing the manuscript; and Pat Struse for preparing the manuscript.

# Contents

# PHYSICS OF CHAOS IN HAMILTONIAN SYSTEMS

*Chapter 1*

# DISCRETE AND CONTINUOUS MODELS

## 1.1 Coexistence of the Dynamical Order and Chaos

Hamiltonian systems are carriers of chaos. With minimal restrictions, the phase space of an arbitrary dynamic Hamiltonian system contains regions where motion is accompanied by a mixing of trajectories in the phase space. The analytical and graphic methods currently available are not good enough to capture the dynamics, which can be either chaotic or regular. More or less successful methods of studying dynamics exist only for systems which do not have more than two degrees of freedom. A system with $N$ degrees of freedom is characterised by $N$ pairs of the generalised co-ordinates $(q_1, \dots, q_N)$ and momentums $(p_1, \dots, p_N)$, which satisfy the Hamiltonian equation of motion

$$\dot{p}_i = -\frac{\partial H}{\partial q_i}, \quad \dot{q}_i = \frac{\partial H}{\partial p_i}, \quad (i = 1, \dots, N) \tag{1.1.1}$$

where

$$H(p, q) \equiv H(p_1, q_1, \dots, p_N, q_N) \tag{1.1.2}$$

is the Hamiltonian of the system. The Hamiltonian depends on time and we will only consider the situation when this dependence is time-periodic with period

$$T = 2\pi/\nu\,. \tag{1.1.3}$$

That is, if $H = H(p, q; t)$, then

$$H(p, q; t + T) = H(p, q, t)\,. \tag{1.1.4}$$

We define system (1.1.4) as having $N+1/2$ degrees of freedom since the time variable is an additional canonical variable.

Chaotic trajectories occur in $N = 1\ 1/2$ degrees of freedom system of the general type. The trajectories of the system can be displayed in a two-dimensional phase plane such that the set of points $(p(t_n), q(t_n))$ correspond to the time instants $t_n = t_0 + nT$. This method of representing a trajectory is known as the Poincaré map. In fact, one sometimes find that

$$(p_{n+1}, q_{n+1}) = \hat{T}(p_n, q_n) \tag{1.1.5}$$

where

$$p_n \equiv p(t_n)\,, \quad q_n \equiv q(t_n)$$

and $\hat{T}$ is a shift operator by time $T$. The Poincaré map (1.1.5) simplifies the original problem (1.1) which consists of an integration of a part of the motion.

Despite the low dimensionality of such systems with chaos, many models have been developed and thus merit discussion. This chapter presents some of the more common types of the map $\hat{T}$. This discussion is also taken up in the following chapters.

Prior to a systematic description of the different models, it is worthwhile to present a view of the plane for a typical system with 1 1/2 degrees of freedom and bounded motion. A schematic picture is presented in Fig. 1.1.1. The dots belong to a trajectory in the domain of chaotic dynamics (stochastic sea) and the curves are closures at the Poincaré map for quasi-periodic (non-chaotic) motion in the so-called islands. In reality, isolated domains of chaotic motion also exist inside the islands. A magnification of an island will show that it resembles the main picture in Fig. 1.1.1, thus confirming the complexity of chaotic dynamics.

The coexistence of both regions of stable dynamics and chaos in the phase space is one of the most striking and wonderful discoveries ever made. It enables one to analyse the onset of chaos and the appearance of the minimal region of chaos. Although not much is known about this field, it is nevertheless evident that a seed of chaos (a so-called stochastic layer) exists which replaces the vicinity of destroyed separatrices.

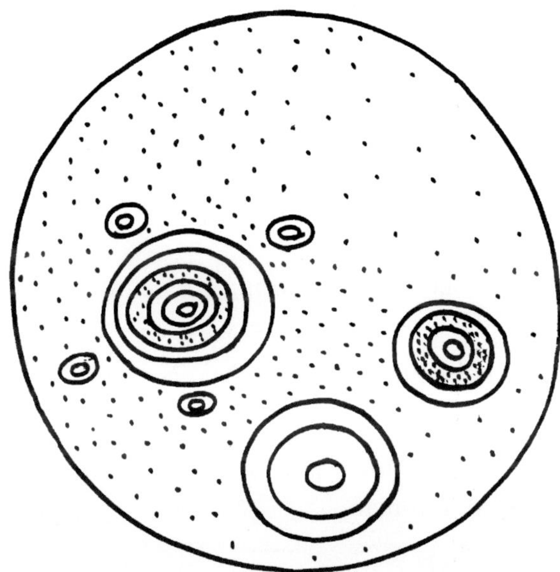

Fig. 1.1.1. Structure of the phase space.

## 1.2 The Standard Map (Kicked Rotator)

This map, also referred to as the Chirikov-Taylor map, emerges in many physical problems. It offers a simplified model of the onset of chaos by retaining the typical complex features of the problem.

The Hamiltonian of a system which can be described by the standard map is expressed as

$$H = \frac{1}{2}I^2 - K\cos\theta \sum_{n=-\infty}^{\infty} \delta\left(\frac{t}{T} - n\right). \qquad (1.2.1)$$

It defines the system with an unperturbed Hamiltonian,

$$H_0 = \frac{1}{2}I^2, \qquad (1.2.2)$$

which is affected by a periodic sequence of kicks ($\delta$-pulses) with the period

$$T = 2\pi/\nu. \qquad (1.2.3)$$

Expression (1.2.2) corresponds either to a particle's free motion, $I = p$, $\theta = x$, or to a free rotator when the variable $\theta$ is cyclic, that is, $\theta \in (0, 2\pi)$. Let us consider the latter case. By using the identity

$$\sum_{n=-\infty}^{\infty} \delta\left(\frac{t}{T} - n\right) = \sum_{n=-\infty}^{\infty} \cos\left(2\pi n \frac{t}{T}\right), \tag{1.2.4}$$

we rewrite (1.2.1) in the following form:

$$H = \frac{1}{2} I^2 - K \cos\theta \sum_{m=-\infty}^{\infty} \cos m\nu t\,. \tag{1.2.5}$$

The Hamiltonian (1.2.5) can also be considered as a particular case of a more general Hamiltonian,

$$H = \frac{1}{2} p^2 + \sum_m V_m \cos(k_m x - \omega_m t)\,, \tag{1.2.6}$$

when

$$k_m = 1(\forall m)\,, \quad V_m = V_0 = -K(\forall m)\,, \quad \omega_m = m\nu$$

and the sum in (1.2.6) is performed over $m \in (-\infty, \infty)$. Thus, the Hamiltonian (1.2.1) corresponds to a particle motion in a periodic wave packet with an infinite number of harmonics of equal amplitude.

The equations of motion derived from (1.2.1) have the form

$$\begin{aligned} \dot{I} &= -K \sin\theta \sum_{n=-\infty}^{\infty} \delta\left(\frac{t}{T} - n\right) \\ \dot{\theta} &= I\,. \end{aligned} \tag{1.2.7}$$

Between two $\delta$-functions, $I = \text{const}$ and $\theta = It + \text{const}$. At each step, or kick, represented by the $\delta$-function, the variable $\theta$ remains continuous and the action $I$ changes by the value $-K\sin\theta$, which can be obtained by integrating (1.2.7) with respect to time in a small vicinity of the $\delta$-function. If we assume that $(I, \theta)$ are the values of the variables just before the $n$-th kick, and that $(\bar{I}, \bar{\theta})$ are the same values before the next $(n+1)$-th kick, it follows that the map, derived from (1.2.7), is

$$\begin{aligned} \bar{I} &= I - K \sin\theta \\ \bar{\theta} &= \theta + \bar{I}\,, \end{aligned} \tag{1.2.8}$$

which is equivalent to the equations of motion in (1.2.7). Here we shall restrict ourselves to some brief comments on the map (1.2.8) by postponing the main discussion to Chapter 2.[1]

For $K = 0$, there is no perturbation and the solution of (1.2.8) is trivial:

$$I_n = \text{const.} = I_0\,, \quad \theta_n = \theta_0 + nI_0\,. \tag{1.2.9}$$

It describes the straight line on the phase plane $(I, \theta)$ as shown in Fig. 1.2.1(a). The same trajectory can be put either on the torus $I(\text{mod } 2\pi)$, $\theta(\text{mod } 2\pi)$, as shown in Fig. 1.2.1(b), or on the cylinder $I \in (-\infty, \infty)$, $\theta(\text{mod } 2\pi)$. Both ways are equivalent in terms of presenting a trajectory or a phase portrait of system (1.2.8) and their convenience is the only consideration in making such a choice. This situation arises because of the existence of a fundamental domain, $[I \in (0, 2\pi); \theta \in (0, 2\pi)]$, for map (1.2.8). This domain can be either arbitrarily shifted or multiplied by integers in both directions.

For small values of $K$, one can consider an approximation of (1.2.8) by introducing derivatives instead of finite differences. Hence

$$\frac{dI}{dt} = -\frac{1}{T} K \sin\theta\,, \quad \frac{d\theta}{dt} = \frac{1}{T} I\,, \tag{1.2.10}$$

or

$$\frac{d^2\theta}{d(t/T)^2} + K \sin\theta = 0\,, \tag{1.2.11}$$

[1] The standard map was proposed by J. B. Taylor [Ta 69] as a model to study the existence of the invariants of motion in magnetic traps. B. V. Chirikov used a more formal means to obtain the standard map [C 79]. Later, it became clear that the situation described by the standard model occurs in numerous applications on physical systems [see (LiL 93)].

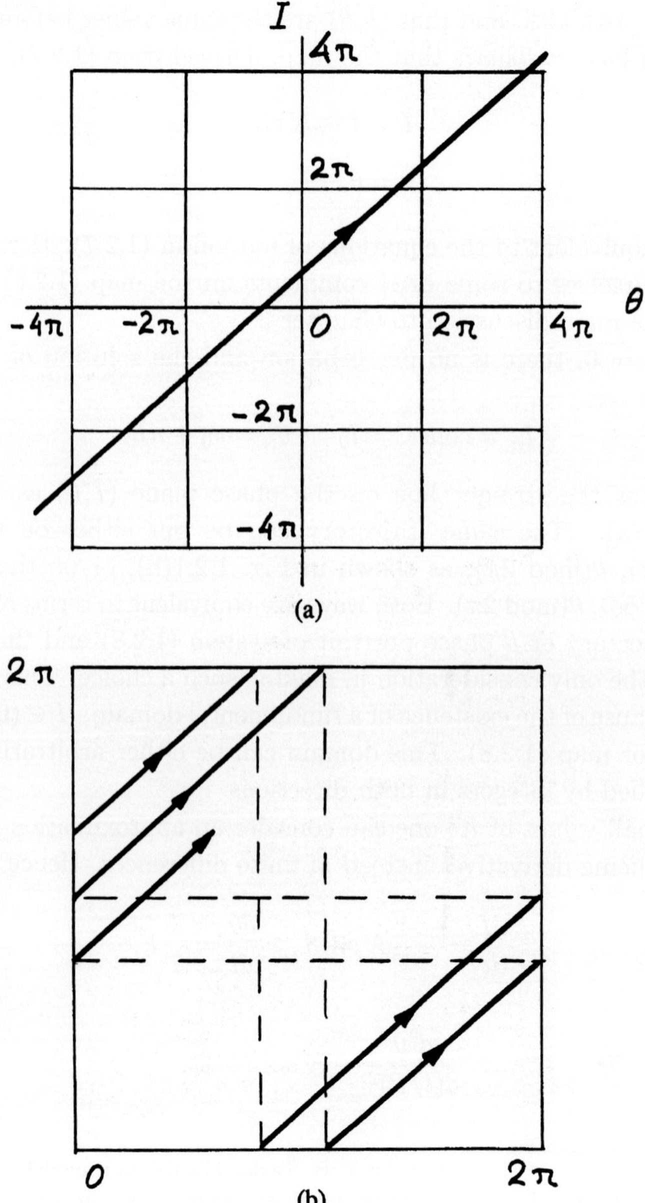


Fig. 1.2.1. A trajectory in: (a) double-periodic phase space; (b) fundamental domain (on the torus).

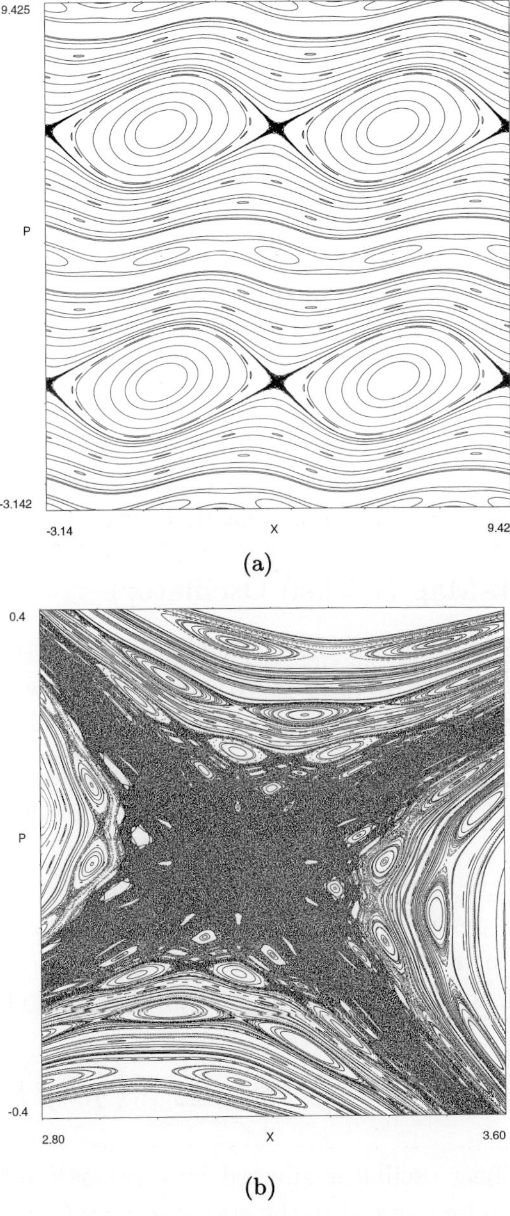


Fig. 1.2.2. The phase portrait of the standard map ($K = 0.5$): (a) part of the phase space $2 \times 2$ periods; (b) magnification of an area near the saddle point.

which is the pendulum equation with Hamiltonian

$$H = \frac{1}{T}\left(\frac{1}{2}I^2 - K\cos\theta\right) \tag{1.2.12}$$

and frequency $K^{1/2}$. Up to a constant, the Hamiltonian (1.2.12) coincides with (1.2.5) if we neglect all the terms in the sum except for $m = 0$. This is a very poor approximation, and it can be seen after a comparison between the integrable pendulum phase portrait and that of the original map (1.2.8), as shown in Fig. 1.2.2. We use four elementary fundamental domains to make Fig. 1.2.2 more representative. Such complexity, as well as the variety of trajectory types, cannot be obtained by any known perturbation method. Narrow zones with chaotic trajectories are called stochastic layers. Inside these stochastic layers are islands with nested curves, sub-islands and smaller stochastic layers.

## 1.3 The Web-Map (Kicked Oscillator)

The web-map is a result of the consideration of particle motion in a constant magnetic field and an electrostatic wave packet propagating perpendicularly to the magnetic field. The Hamiltonian of the system is

$$H = \frac{1}{2}(p^2 + \omega_0^2 x^2) - \frac{\omega_0 K}{T}\cos x \sum_{m=-\infty}^{\infty} \delta\left(\frac{t}{T} - n\right), \tag{1.3.1}$$

where $\omega_0$ is the gyro-frequency and $K$ the dimensionless parameter proportional to the perturbation amplitude. The derivation of the map is similar to that for a standard map.

Consider the equation of motion that follows from (1.3.1):

$$\ddot{x} + \omega_0^2 x = -\frac{\omega_0 K}{T}\cos x \sum_{m=-\infty}^{\infty} \delta\left(\frac{t}{T} - n\right). \tag{1.3.2}$$

It describes a linear oscillator affected by a periodic set of kicks. The model (1.3.2) can be reduced to the standard map (1.2.1) by using only a singular transform: $\omega_0 \to 0$, $K \to \infty$, $\omega_0 K = \text{const}$. For any finite $K$, the results described below cannot be simply considered in the limit

$\omega_0 = 0$, and that makes the kicked-oscillator model independent of the kicked-rotator model. Next, we define

$$x_n = x(t_n - 0)\,, \quad \dot{x}_n = v_n = p_n/m = \dot{x}(t_n - 0) \tag{1.3.3}$$

with the discrete time being $t_n = nT$. From (1.3.2), it follows that, directly by integration,

$$\begin{aligned} x(t_n + 0) &= x(t_n - 0) \\ \dot{x}(t_n + 0) &= \dot{x}(t_n - 0) - K\omega_0 \sin(x_n - \omega t_n)\,. \end{aligned} \tag{1.3.4}$$

Between two adjacent kicks, the solution of (1.3.2) satisfies the free motion equation,

$$\ddot{x} + \omega_0^2 x = 0\,,$$

which permits one to express $x_{n+1} = x(t_{n+1} - 0)$, $\dot{x}_{n+1} = \dot{x}(t_{n+1} - 0)$ through $x(t_n + 0)$, $\dot{x}(t_n + 0)$. After applying (1.3.4), we obtain the map

$$\begin{aligned} u_{n+1} &= (u_n + K \sin v_n) \cos\alpha + v_n \sin\alpha \\ v_{n+1} &= -(u_n + K \sin v_n) \sin\alpha + v_n \cos\alpha \end{aligned} \tag{1.3.5}$$

where the following dimensionless variables are introduced:

$$u = \dot{x}/\omega_0\,, \quad v = -x\,, \quad \alpha = \omega_0 T\,. \tag{1.3.6}$$

The map obtained, (1.3.5), is called the web-map for reasons that will soon become clear. Using the complex variable $z = u + iv$, we rewrite the web-map (1.3.5) as

$$z_{n+1} = \left( z_n + K \sin \frac{z - z^*}{2i} \right) e^{i\alpha}\,, \tag{1.3.7}$$

that is,

$$z_{n+1} = \hat{R}_\alpha (1 + \hat{S}(K)) z_n \tag{1.3.8}$$

where $\hat{R}_\alpha$ is the rotation transform by $\alpha$ and $S(K)$ the shear transform along Im $z$ with intensity parameter $K$.[2]

[2]The web-map was derived in [ZZSUC 86] and used on different physical problems, including the problem of quasi-crystal symmetry generation [see (ZSUC 91)].

The most interesting case involving map (1.3.5) is that of the resonance between sequences of kicks and oscillator frequency $\omega_0$. To simplify this condition, consider

$$\alpha = \omega_0 T = 2\pi/q \tag{1.3.9}$$

with an integer $q$ and denote

$$\hat{M}_q = \hat{R}_{2\pi/q} \cdot \hat{S}(K)\,. \tag{1.3.10}$$

Some simple examples of the resonance case (1.3.9) include

$$\begin{aligned} &\hat{M}_1\colon u_{n+1} = u_n + K \sin v_n\,, \qquad v_{n+1} = v_n \\ &\hat{M}_2\colon u_{n+1} = -u_n - K \sin v_n\,, \quad v_{n+1} = -v_n \\ &\hat{M}_2^2\colon u_{n+2} = u_n + 2K \sin v_n\,, \quad v_{n+2} = v_n \end{aligned} \tag{1.3.11}$$

which correspond to the operating regimes of the earliest cyclotrons. A more complicated case arises when $q = 4$:

$$\hat{M}_4\colon u_{n+1} = v_n \quad v_{n+1} = -u_n - K \sin v_n\,. \tag{1.3.12}$$

It is non-integrable and possesses chaotic trajectories (see Fig. 1.3.1). For small $K \ll 1$ in the first approximation, it is easy to derive

$$\hat{M}_4^4 \approx \hat{M}_H\colon \begin{array}{l} \bar{u} = u + 2K \sin \bar{v} \\ \bar{v} = v - 2K \sin u \end{array}, \tag{1.3.13}$$

where $\hat{M}_H$ corresponds to the so-called Harper kicked oscillator described by the Hamiltonian

$$H_H = -\frac{1}{2} K \left\{ \cos v + \cos u \sum_{n=-\infty}^{\infty} \delta\left( \frac{1}{4} \frac{t}{T} - n \right) \right\}.^3 \tag{1.3.14}$$

For $K = 0$, the map (1.3.5), or (1.3.7), describes a pure rotation as opposed to the standard map (1.2.8). This renders both maps quite

[3] It is shown in [D 95] that an exact relationship exists between $\hat{M}_4$ and $\hat{M}_H$.

different from each other. To view it, rewrite the Hamiltonian (1.3.1) by using the action-angle variables:

$$P = (2I\omega_0)^{1/2}\cos\phi\,, \quad x = (2I/\omega_0)^{1/2}\sin\phi\,. \tag{1.3.15}$$

After substituting (1.3.15) in (1.3.1), one derives

$$H = \omega_0 I - \frac{\omega_0}{T} K \cos[(2I/\omega_0)^{1/2}\sin\phi] \sum_{n=-\infty}^{n+\infty} \delta\left(\frac{t}{T} - n\right). \tag{1.3.16}$$

The unperturbed Hamiltonian in (1.2.1) is $H_0 = \frac{1}{2}I^2$ and the corresponding frequency satisfies the condition

$$\frac{d\omega(I)}{dI} = \frac{d^2 H_0}{dI^2} \neq 0 \tag{1.3.17}$$

while in the case of (1.3.16), $H_0 = \omega_0 I$ and

$$\frac{d\omega(I)}{dI} = 0\,, \tag{1.3.18}$$

which indicates the existence of degeneracy. Hence the standard map and the web-map complement each other in various physical situations.

Another important property of the web-map (1.3.5) is that it can be considered a dynamical generator of the $q$-fold symmetry for the resonance condition (1.3.9), which is of the crystalline type for

$$q \in \{q_c\} = \{1, 2, 3, 4, 6\} \tag{1.3.19}$$

and the quasi-crystal type for $q \notin \{q_c\}$. This topic will be discussed in greater detail in Chapter 8. A fundamental domain exists for (1.3.19). Examples of a phase plane with narrow stochastic layers are given in Fig. 1.3.1 for $q = 4$ and Fig. 1.3.2 for $q = 3$ and 6. These illustrations demonstrate the existence of the stochastic web of a corresponding symmetry in the phase plane. The stochastic web is a net with each part being a stochastic layer. One can say that the net of channels, which constitute the stochastic web, provides particle transport along the web. The stochastic web can be finite or infinite. The issues related to it will be discussed in Chapter 8.

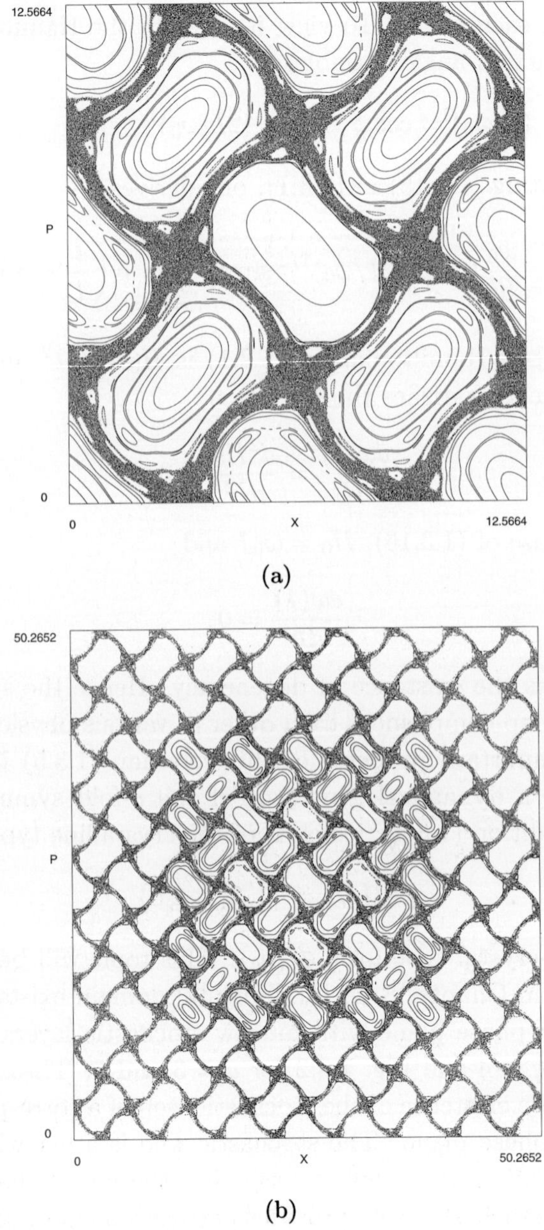


Fig. 1.3.1. Stochastic web and invariant curves for the four-fold symmetry ($q = 4$, $K = 1.5$): (a) element of the web; (b) magnification of the central part of the web.

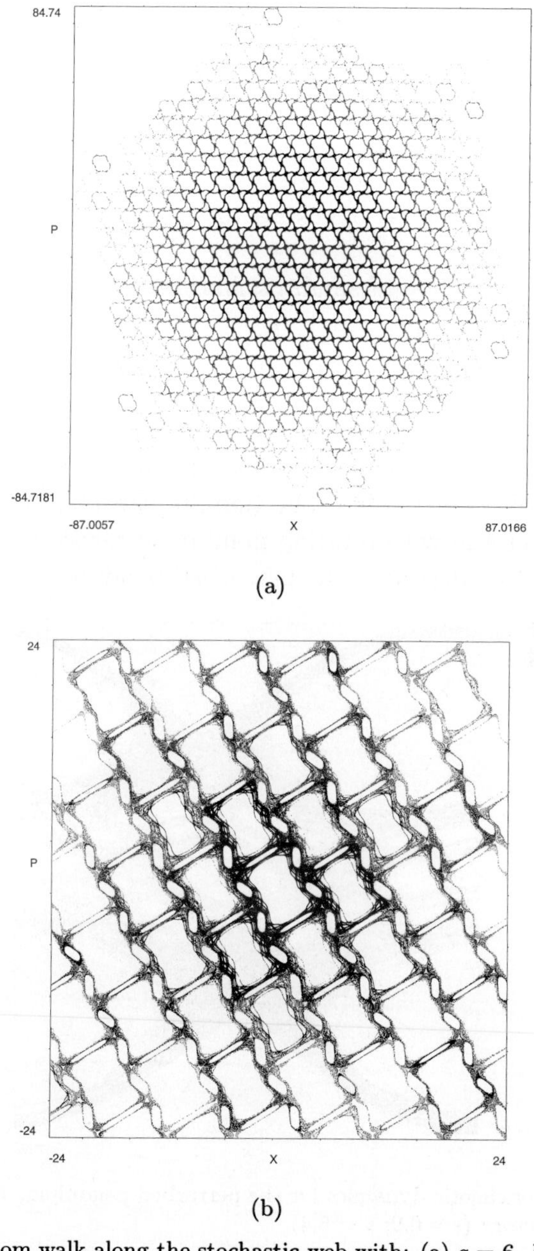

(a)

(b)

Fig. 1.3.2. Random walk along the stochastic web with: (a) $q = 6$, $K = 1.2$; (b) $q = 3$, $K = 1.7$.

## 1.4 Perturbed Pendulum

The perturbed pendulum is a typical model of the continuous equation of motion (no kicks) where one can still introduce a map. The Hamiltonian of the model is

$$H = \frac{1}{2}\dot{x}^2 - \omega_0^2 \cos x + \epsilon\omega_0^2 \cos(kx - \nu t)\,. \tag{1.4.1}$$

Here $\omega_0$ is the frequency of small oscillations in the unperturbed pendulum with Hamiltonian

$$H_0 = \frac{1}{2}\dot{x}^2 - \omega_0^2 \cos x\,, \tag{1.4.2}$$

where $\epsilon$ is the small dimensionless parameter of perturbation and $\nu$ the frequency of perturbation. The Hamiltonian (1.4.1) corresponds to the pendulum (1.4.2) with rotating point of suspension.[4] The following equation of motion describes the perturbed pendulum:

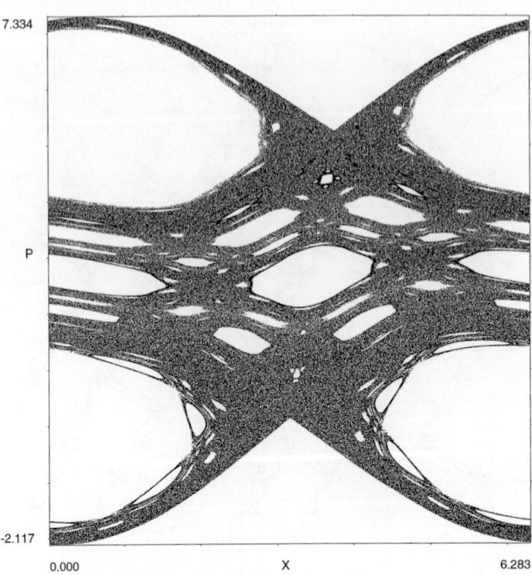


Fig. 1.4.1. Area of chaotic dynamics for the perturbed pendulum. All points belong to the only trajectory ($\epsilon = 0.9$; $\nu = 5.4$).

---

[4]See [LL 76] for the corresponding derivation of the Lagrangian in the model.

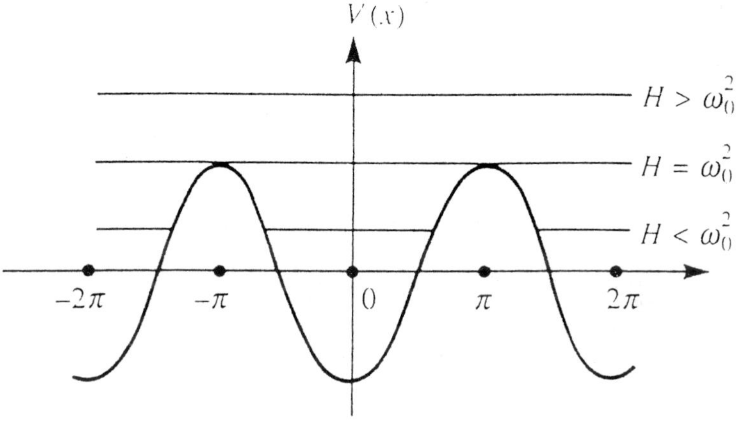

(a)

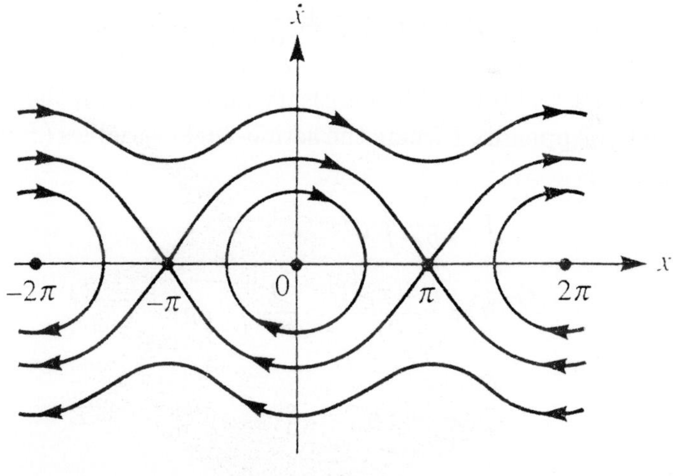

(b)

Fig. 1.4.2. Unperturbed pendulum: (a) periodic potential; (b) the phase portrait.

$$\ddot{x} + \omega_0^2 \sin x = \epsilon k \omega_0^2 \sin(kx - \nu t)\,. \tag{1.4.3}$$

The Poincaré map can be considered as a set of trajectory points $(x, \dot{x})$ on the phase plane taken at each period $T = 2\pi/\nu$ of perturbation. Figure 1.4.1 exemplifies the complexity of the corresponding dynamics.

The problem of (1.4.3) will be considered in Chapter 2. A good knowledge of unperturbed pendulum dynamics is required. The phase portrait of the pendulum (1.4.2) is shown in Fig. 1.4.2. Its main feature is the existence of a separatrix at the energy

$$H_0 = H_s = \omega_0^2 \tag{1.4.4}$$

with a corresponding solution for the co-ordinate

$$x = 4 \arctan \exp(\pm \omega_0 t) - \pi \tag{1.4.5}$$

(it has a form called a "kink") and for the momentum $p$ or velocity if the mass is equal to one,

$$p = v = \dot{x} = \pm \frac{2\omega_0}{\cosh \omega_0 t}, \tag{1.4.6}$$

which is called a "soliton". The necessary information on the pendulum is presented in Appendix 1 where the action-angle variables $(I, \theta)$ defined by the relations

$$\begin{aligned} I &= \frac{1}{2\pi} \oint p dx \\ \theta &= \frac{\partial S(x, I)}{\partial I} = \frac{\partial}{\partial I} \int^x dp\, x \end{aligned} \tag{1.4.7}$$

with

$$p = \pm [2(H_0 + \omega_0^2 \cos x)]^{1/2} \tag{1.4.8}$$

are used.

The integration in the definition of $I$ is performed over the period of motion $2\pi/\omega(I)$ where the nonlinear frequency

$$\omega(I) = dH_0(I)/dI \tag{1.4.9}$$

is introduced and the condition $H_0 < \omega_0^2$ is applied. In the case of $H_0 \geq \omega_0^2$, the integration in (1.4.7) is performed over the interval $x \in (0, 2\pi)$ for both positive and negative $p$. The details are found in Appendix 1.

Instead of using $\omega(I)$, it is sometimes convenient to use $\omega(H_0)$ and the connection, $I = I(H_0)$, which follows from (1.4.7).

As will be clearly seen in Chapter 2, the time dependence of the velocity $\dot{x}$ in the vicinity of the unperturbed separatrix is crucial to the problem (1.4.1) (see also (A.1.11)). This dependence is shown in Fig. 1.4.3 for $H_0 < H_s$. It resembles a periodic set of kicks that makes it possible to define the so-called separatrix map, the definition of which can be found in Section 2.3. The importance of the perturbed pendulum model (1.4.1) lies in its universality in studying the separatrix destruction.

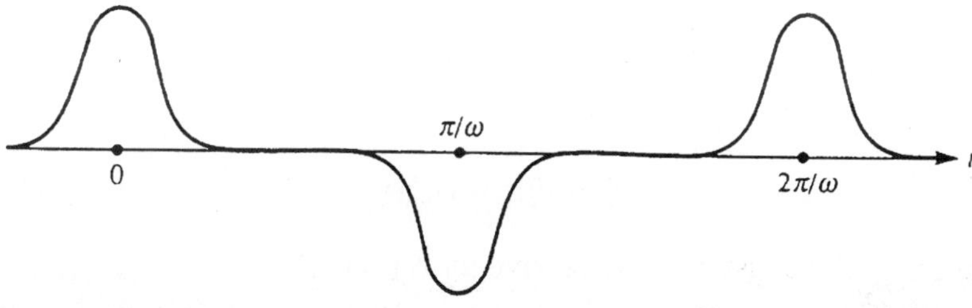


Fig. 1.4.3. The pendulum velocity in the vicinity of separatrix as a function of time.

## 1.5 Perturbed Oscillator

For fairly small amplitudes of oscillations, the pendulum equation can be reduced to the linear oscillator equation and the problem (1.4.1) replaced by the perturbed oscillator problem with Hamiltonian

$$H = \frac{1}{2}(\dot{x}^2 + \omega_0^2 x^2) + \epsilon \frac{\omega_0^2}{k^2} \cos(kx - \nu t) \tag{1.5.1}$$

and the equation of motion

$$\ddot{x} + \omega_0^2 x = \frac{1}{k} \epsilon \omega_0^2 \sin(kx - \nu t) \,. \tag{1.5.2}$$

Problem (1.5.2) has another physical interpretation. It is equivalent to the motion of a particle in a constant magnetic field, $B_0$, and in the field of a plane wave travelling perpendicularly to the magnetic field. In this case, the equation of motion of a particle is written as

$$\ddot{\mathbf{r}} = \frac{e}{mc}[\dot{\mathbf{r}}, \mathbf{B}_0] + \frac{e}{m}\mathbf{E}_0 \sin(\mathbf{kr} - \nu t)\,, \tag{1.5.3}$$

where $[,]$ means vector product. Assuming that $\mathbf{B}_0$ is directed along the $z$-axis, the vector $\mathbf{r}$ lies in the plane $(x, y)$, and vectors $\mathbf{k}$ and $\mathbf{E}_0$ are directed along the $x$-axis (a longitudinal wave), it follows from (1.5.3) that

$$\begin{aligned} \ddot{x} &= \omega_0 \dot{y} + \frac{1}{k}\epsilon\omega_0^2 \sin(kx - \nu t)\,, \\ \ddot{y} &= -\omega_0 \dot{x}\,, \end{aligned} \tag{1.5.4}$$

where

$$\omega_0 = eB_0/mc\,, \quad \epsilon\omega_0^2 = eE_0 k/m\,. \tag{1.5.5}$$

From (1.5.4), it follows that the existence of an integral of motion is

$$\dot{y} + \omega_0 x = \text{const}\,. \tag{1.5.6}$$

Assuming that const $= 0$, we arrive at Eq. (1.5.2).

System (1.5.1) possesses the same degeneracy property in its unperturbed part as the kicked-pendulum case (1.3.1). As in (1.3.15), we can introduce the action-angle variables $(I, \phi)$:

$$\begin{aligned} \dot{x} &= (2\omega_0 I)^{1/2} \cos\phi\,, \\ x &= (2I/\omega_0)^{1/2} \sin\phi\,. \end{aligned} \tag{1.5.7}$$

Expressed via the following variables, the Hamiltonian (1.5.1) is

$$\begin{aligned} H &= \omega_0 I + \epsilon V(I, \phi; t)\,, \\ V(I, \phi; t) &= \frac{1}{k^2}\omega_0^2 \cos\left[k\left(\frac{2I}{\omega_0}\right)^{1/2} \sin\phi - \nu t\right]. \end{aligned} \tag{1.5.8}$$

The unperturbed part of the Hamiltonian, $H_0 = \omega_0 I$, does not satisfy the non-degeneracy condition (1.3.17). Therefore, in the case of a resonance

$$n\omega_0 = \nu\,, \tag{1.5.9}$$

where $n$ is an integer, the amplitude of the oscillator increases strongly.

Chapter 2 will demonstrate that the degeneracy of the unperturbed Hamiltonian causes a stochastic web to be generated in the phase space of a system when affected by a periodic perturbation.

## 1.6 Billiards

A system involving a point-size ball bouncing between different scatterers with elastic collisions is called a billiard. An example of a Sinai

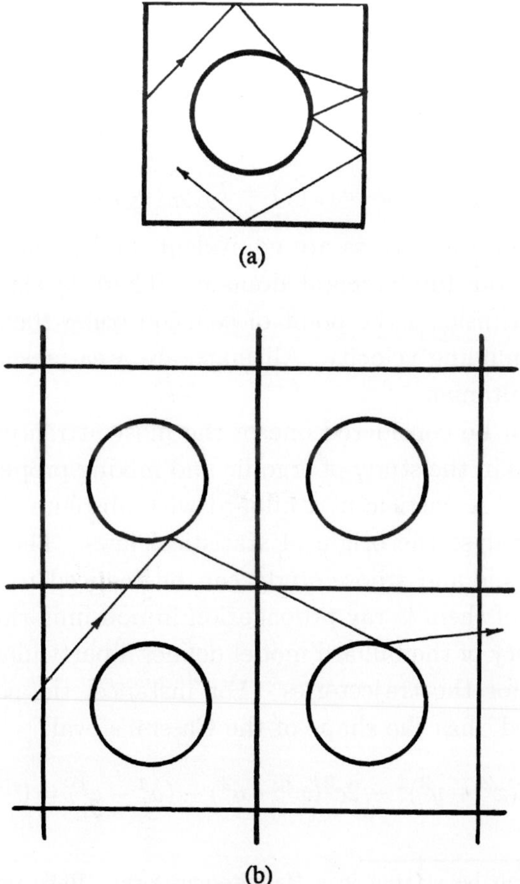


Fig. 1.6.1. Sinai billiard (a) and its double-periodic continuation (b).

billiard is shown in Fig. 1.6.1(a). The figure can be considered a fundamental domain and it can be periodically continued in $x$-, $y$- or in both directions, as shown in Fig. 1.6.1(b). In the latter case, it is called Lorentz gas. The dynamics of a particle (ball) in a two-dimensional billiard can be described using two canonical pairs of variables $(x, v_x; y, v_y)$. In reality, a conservation law,

$$v_x^2 + v_y^2 = \text{const} = 1\,, \tag{1.6.1}$$

exists due to the elasticity of collisions. This property makes it possible to use a two-dimensional map linking one collision to the next:

$$(x_{n+1}, v_{x,n+1}) = \hat{T}_x(x_n, v_{x,n}) \tag{1.6.2}$$

or

$$(y_{n+1}, v_{y,n+1}) = \hat{T}_y(y_n, v_{y,n})\,. \tag{1.6.3}$$

Both of the above equations are equivalent, and points $x$ or $y$ can be obtained from the fundamental domain. There is also another type of map which considers the point of collision using the inner scatterer and its corresponding velocity. All maps are area-preserving since the system is Hamiltonian.

Billiards can be considered one of the most attractive types of dynamical models in the study of ergodic and mixing properties in Hamiltonian systems. A particle in a billiard with absolute elastic collisions was used to analyse the origin of statistical laws. There are different physical problems and whose study can be reduced to a billiard-type problem. One of them is ray propagation in non-uniform waveguides.[5]

The geometry of the billiard model defines a particular type of chaos and an order for the trajectories. For instance, the scatterer in the "Cassini billiard" has the shape of the Cassini's oval:

$$(x^2 + y^2)^2 - 2c^2(x^2 - y^2) - (a^4 - c^4) = 0\,. \tag{1.6.4}$$

[5]Ray dynamics can be written in a Hamiltonian form. Reflecting rays from the waveguide walls render the problem similar to the billiard problem. One can find out more on ray chaos in [AZ 91] and [Ab 91].

The shape of the oval is sensitive to the parameters $a$ and $c$ and can have both concave and convex parts. The phase space of the Cassini billiard is much more complicated than that of the Sinai billiard since it permits a non-ergodic motion due to the islands in the phase space. One can also consider an analog of the Lorentz gas if a double-periodic continuation of the scatterers is made (see Fig. 1.6.2). The concave part of the scatterer's boundary makes it possible to trap the trajectories for an arbitrarily long (but finite) time.

The specific properties of the billiards are discussed in Sections 7.3 and 7.4.

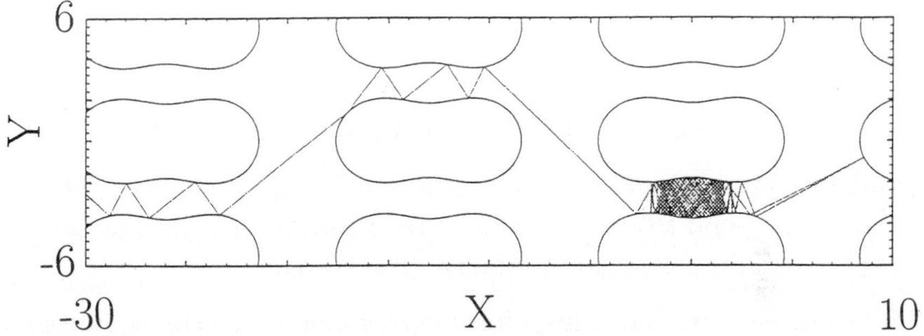


Fig. 1.6.2. Example of a trajectory in the Cassini billiard (with double-periodic continuation).

## Conclusions

1. One can select a special set of time instants, $\{t_j\} = \{t_1, t_2, \ldots\}$, and find the dependence of system co-ordinates and moments, ($\mathbf{p}_{n+1}$, $\mathbf{q}_{n+1}$), at time $t_{n+1}$ as a function of the co-ordinates and moments $(\mathbf{p}_n, \mathbf{q}_n)$ at time $t_n$. The corresponding time shift operator $\hat{T}_n$

$$\hat{M}: (\mathbf{p}_{n+1}, \mathbf{x}_{n+1}) = \hat{T}_n(\mathbf{p}_n, \mathbf{x}_n)$$

defines the Poincaré map $\hat{M}$. The conveniently taken set, $\{t_j\}$, can help, that is, the investigation of the properties of the map $\hat{M}$ is easier than a study of the initial equation of motion. Any map $\hat{M}$ is area-preserving for Hamiltonian systems (it is equivalent to the Liouville theorem on the phase volume conservation).

2. There are few models of maps which are carriers of typical properties of physical systems. The most popular model is the standard map,

$$\hat{M}_{\text{st}}: \begin{array}{l} p_{n+1} = p_n + K \sin x_n \\ x_{n+1} = x_n + p_{n+1} \,, \end{array}$$

which corresponds to a periodically kicked rotator. An alternative to map $\hat{M}_s$, with respect to some important condition of degeneracy, is the web-map,

$$\hat{M}_{\alpha}: \begin{array}{l} u_{n+1} = (u_n + K \sin v_n) \cos \alpha + v_n \sin \alpha \\ v_{n+1} = -(u_n + K \sin v_n) \sin \alpha + v_n \cos \alpha \,, \end{array}$$

which generates stochastic webs for $\alpha = 2\pi/q$ with integer $q > 2$. The map $\hat{M}_\alpha$ corresponds to a periodically kicked oscillator.

3. Billiard-type systems naturally generate a map to connect variables that correspond to two continuous reflections of a ball. The set $\{t_j\}$ is not periodic for maps in billiard-type systems.

4. It is not easy to find a map-form corresponding to the equations of motion for typical physical systems, such as the perturbed oscillator or pendulum. We are lucky when that happens.

*Chapter 2*

# SEPARATRIX CHAOS

## 2.1 Nonlinear Resonance and Chain of Islands

This section seeks to provide a qualitative understanding of the chaotic dynamics resulting from a complexity of motion which arises due to the interaction of resonances. Here we describe the structure of the resonances.

The perturbed motion of a system with one degree of freedom by Hamiltonian in action-angle variables is

$$H = H_0(I) + \epsilon V(I, \theta, t)\,, \tag{2.1.1}$$

where $\epsilon \ll 1$ is a small dimensionless parameter of perturbation and $V$ is periodic in time with the period $T = 2\pi/\nu$. Therefore, $V$ can be expanded in a double Fourier series in $\theta$ and $t$:

$$\begin{aligned} V(I, \theta, t) &= \frac{1}{2} i\epsilon \sum_{k,\ell} V_{k,\ell}(I) e^{i(k\theta - \ell\nu t)} + \text{c.c.} \\ V^*_{k,\ell} &= V_{-k,-\ell}\,, \end{aligned} \tag{2.1.2}$$

where c.c. designates the complex conjugate terms.

Using expressions (2.1.1) and (2.1.2), the canonical equations of motion can be transformed into

$$\dot{I} = -\epsilon \frac{\partial V}{\partial \theta} = -\frac{1}{2} \sum_{k,\ell} k V_{k,\ell}(I) e^{i(k\theta - \ell \nu t)} + \text{c.c.}$$

$$\dot{\theta} = \frac{\partial H}{\partial I} = \frac{dH_0}{dI} + \epsilon \frac{\partial V(I,\theta,t)}{\partial I} = \omega(I) + \frac{1}{2}\epsilon \sum_{k,\ell} \frac{\partial V_{k,\ell}(I)}{\partial I} e^{i(k\theta - \ell\nu t)} + \text{c.c.}\,, \tag{2.1.3}$$

where we introduce the frequency of oscillations for unperturbed motion:

$$\omega(I) = \frac{dH_0}{dI}\,. \tag{2.1.4}$$

The resonance condition implies that the following equation must be satisfied:

$$k\omega(I) - \ell\nu = 0 \tag{2.1.5}$$

where $k$ and $\ell$ are integers. This means that we have to specify a pair of integers, $(k_0, \ell_0)$, and the corresponding value, $I_0$, such that (2.1.5) converts into the following identity:

$$k_0 \omega(I_0) = \ell_0 \nu\,. \tag{2.1.6}$$

As a rule, the values $(k_0, \ell_0; I_0)$ can be found in abundance. This is due largely to the system's nonlinearity, that is, to the dependence of $\omega(I)$ on $I$.

The way to deal with Eqs. (2.1.3) is to analyse certain simplified situations. First, we examine an isolated resonance (2.1.6) and ignore all other possible resonances. This means that in the equations of motion (2.1.3), only terms with $k = \pm k_0$, $\ell = \pm \ell_0$ should be retained which satisfy the resonance condition at $I = I_0$:

$$\dot{I} = \epsilon k_0 V_0 \sin(k_0\theta - \ell_0 \nu t + \phi)$$

$$\dot{\theta} = \omega(I) + \epsilon \frac{\partial V_0}{\partial I} \cos(k_0\theta - \ell_0\nu t + \phi) \tag{2.1.7}$$

where we set

$$V_{k_0,\ell_0} = |V_{k_0,\,\ell_0}| e^{i\phi} = V_0 e^{i\phi}\,. \tag{2.1.8}$$

It is also assumed that the value

$$\Delta I = I - I_0 \tag{2.1.9}$$

is small, that is, Eq. (2.1.6) is examined in the vicinity of the resonance value of the action $I_0$.

A list of the typical approximations is as follows:

(i) Set $V_0 = V_0(I_0)$ on the right-hand sides of Eqs. (2.1.7);

(ii) Expand the frequency $\omega(I)$ using (2.1.9):

$$\omega(I) = \omega_0 + \omega' \Delta I \tag{2.1.10}$$

where

$$\omega_0 = \omega(I_0)\,, \quad \omega' = d\omega(I_0)/dI\,; \tag{2.1.11}$$

(iii) Neglect the second term of the order of $\epsilon$ in the second equation of Eq. (2.1.7) for frequency.

Finally, we reduce system (2.1.7) to a simplified version:

$$\begin{aligned} \frac{d}{dt}(\Delta I) &= -\epsilon k_0 V_0 \sin\psi \\ \frac{d}{dt}\psi &= k_0 \omega' \Delta I \end{aligned} \tag{2.1.12}$$

where a new phase is introduced:

$$\psi = k_0\theta - \ell_0 \nu t + \phi - \pi\,. \tag{2.1.13}$$

The set of equations in (2.1.12) can be presented in a Hamiltonian form:

$$\frac{d}{dt}(\Delta I) = -\frac{\partial \bar{H}}{\partial \psi}\,; \quad \frac{d}{dt}\psi = \frac{\partial \bar{H}}{\partial(\Delta I)} \tag{2.1.14}$$

where

$$\bar{H} = \frac{1}{2} k_0 \omega' (\Delta I)^2 - \epsilon k_0 V_0 \cos\psi\,. \tag{2.1.15}$$

Expression (2.1.15) is an effective Hamiltonian describing the dynamics of a system in the neighbourhood of the resonance. Variables $(\Delta I, \psi)$ form a canonically conjugate pair.

A comparison of expressions (2.1.15) and (1.4.2) shows that the Hamiltonian $\bar{H}$ describes the oscillations of a nonlinear pendulum. It follows from (2.1.12) that

$$\ddot{\psi} + \Omega_0^2 \sin \psi = 0 \,, \qquad (2.1.16)$$

where the "small amplitude" oscillation frequency is

$$\Omega_0 = (\epsilon k_0^2 V_0 |\omega'|)^{1/2} \,. \qquad (2.1.17)$$

The value $\Omega_0$ is also the frequency of phase oscillations. The formulas obtained in Section 1.4 and Appendix 1 for the nonlinear pendulum can be automatically applied to Hamiltonian (2.1.15).

On the phase plane $(p, x)$, a phase curve defined by the action $I_0$ (see Fig. 2.1.1) corresponds to an exact resonance. The transform (2.1.13), from the polar angle $\theta$ to phase $\psi$, corresponds to the transform in a co-ordinate system rotating at frequency $\ell_0 \nu$. Two examples of the phase portrait in nonlinear resonance in the rotating co-ordinate system are shown in Figs. 2.1.1 and 2.1.2. It follows that for the resonance with an order $k_0$, there is a chain of islands with $k_0$ separatrix cells, $k_0$ hyperbolic singular points and $k_0$ elliptic cells. Thus, the topology of the phase space is changed in the vicinity of the resonance value $I_0$.

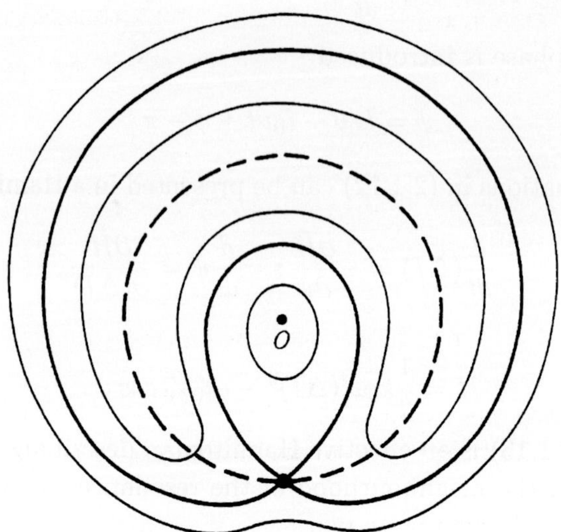


Fig. 2.1.1. The nonlinear resonance ($k_0 = \ell_0 = 1$) on the phase plane $(p, q)$. The dashed line is the unperturbed trajectory with $I = I_0$. The thick line is a new separatrix of the phase oscillations.

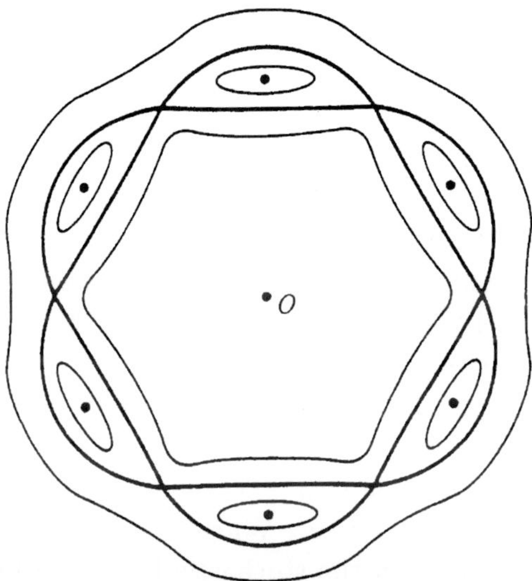


Fig. 2.1.2. Another version of the nonlinear resonance ($k_0 = 6,\ \ell_0 = 1$).

We now turn our attention to determine the conditions under which the system is "trapped" by a nonlinear resonance. The dimensionless parameter $\alpha$ characterises the degree of nonlinearity of oscillations:

$$\alpha = \frac{I_0}{\omega_0}\left|\frac{d\omega(I_0)}{dI}\right| \equiv \frac{I_0}{\omega_0}|\omega'| \,. \tag{2.1.18}$$

The three approximations mentioned above are valid if the following conditions are satisfied:

$$\epsilon \ll \alpha \ll 1/\epsilon \,. \tag{2.1.19}$$

The first inequality means that the nonlinearity must be fairly strong while the second inequality restricts the time of the approximation.

To determine the validity of the Hamiltonian (2.1.15), we need to estimate the amplitude of the phase oscillations:

$$\frac{\max \Delta I}{I_0} \sim \left(\epsilon \frac{V_0}{|\omega'|}\right)^{1/2} \frac{1}{I_0} \sim \left(\frac{\epsilon}{\alpha}\right)^{1/2} \tag{2.1.20}$$

where we put $V_0 \sim H_0 \sim \omega_0 I_0$. In the same way, we derive the frequency width of the resonance from definition (2.1.17):

$$\frac{\max \Delta\omega}{\omega_0} = \frac{\Omega_0}{\omega_0} = (\epsilon\alpha)^{1/2}. \tag{2.1.21}$$

Expression (2.1.20) provides the meaning for the first inequality in (2.1.19) and expression (2.1.21) the meaning for the second, that is, the relative changes in action and frequency attributed to the nonlinear resonance must be small.

We can now easily describe part of the trajectories presented in Figs. 1.2.2(a) and 1.2.2(b). Inside a separatrix loop, there is a set of nested invariant curves which do not correspond to any visible resonant condition. There is also a chain of islands that corresponds to a high value of $\ell_0$ resonance. This resonance chain confines the central elliptic point although there are many that do not do so. Every chain of islands are separated by invariant curves. When chaos occurs, remnants of the resonance islands persist in the stochastic layer (see Fig. 1.2.2(b)) but the islands are not separated by invariant curves.

The pattern described is similar to those found in other models (see Figs. 1.3.1 and 1.4.2). In fact, the pattern is more complicated not only because the island chains are separated by invariant curves, but each island is bordered by a thin stochastic layer. Inside the stochastic layer, remnants of islands are embedded in the domain of chaotic motion and the same pattern is repeated inside each island.

## 2.2 Overlapping of Resonances

A glance at Fig. 1.2.2(a) will reveal that for some situations, one can introduce a parameter,

$$K_c = \frac{\Delta I}{\delta I}, \tag{2.2.1}$$

where $\Delta I$ is the maximum width of a chain of islands and $\delta I$ is the distance between two neighbouring chains. Parameter $K_c$ is meaningful when $K_c \ll 1$, that is, the chains of islands are well separated from each other by a layer of invariant curves. The condition

$$K_c \gtrsim 1 \tag{2.2.2}$$

was proposed by Chirikov as a qualitative criterion in the occurrence of stronger chaos[1] than the one which already exists. It is understood that any perturbation which affects a chain of islands will destroy the separatrices and create stochastic layers instead. This phenomenon is considered in the next section.

For small $\Delta I$ and $\delta I$, we can write them as

$$\Delta\omega = |\omega'|\Delta I\,, \quad \delta\omega = |\omega'|\delta I \tag{2.2.3}$$

and

$$K_c = \frac{\Delta\omega}{\delta\omega} \gtrsim 1\,. \tag{2.2.4}$$

It is easy to estimate the value of $K_c$ from the resonance condition (2.1.6) in the following two cases:

(i) Consider $\nu \ll \omega_0 = \omega(I_0)$ and, for the sake of simplicity, put $k_0 = 1$. The resonance besides (2.1.6) would then satisfy the condition

$$\omega(I_0 + \delta I) = (\ell_0 \pm 1)\nu \tag{2.2.5}$$

or

$$\delta\omega = |\omega(I_0 + \delta I) - \omega(I_0)| = \nu\,. \tag{2.2.6}$$

Thus, we derive

$$K_c = \frac{\Delta\omega}{\nu} = (\epsilon\alpha)^{1/2}\frac{\omega_0}{\nu} \gtrsim 1 \tag{2.2.7}$$

from Eqs. (2.2.4), (2.2.6) and (2.1.21). Since for (2.1.19) $\epsilon\alpha \ll 1$ and $\omega_0/\nu \gg 1$, expression (2.2.7) is the product of a small and large parameter, that is, condition (2.2.7) is non-trivial.

[1]The overlapping criterion was proposed by B. Chirikov fairly early [C 59]. Later, it was studied in various ways, including simulation [for a review, see (C 79, LiL 93)]. The criterion enables one to choose the correct resonances. It is very convenient since it allows a quick estimation of the physical conditions under which the system becomes stochastically unstable to be made.

(ii) Consider the reverse when $\nu \gg \omega_0$ and, for the sake of simplicity, $\ell_0 = 1$. The resonance beside (2.1.6) would then satisfy the condition

$$(k_0 \pm 1)\omega(I_0 + \delta I) = \nu \tag{2.2.8}$$

or

$$\delta\omega = |\omega(I_0 + \delta I) - \omega(I_0)| = \left|\frac{\nu}{k_0} - \frac{\nu}{k_0 \pm 1}\right| \sim \frac{\omega_0^2}{\nu}. \tag{2.2.9}$$

Once again, by applying (2.2.9) and (2.1.21) in (2.2.1), we obtain

$$K_c = (\epsilon\alpha)^{1/2}\frac{\nu}{\omega_0} \gtrsim 1. \tag{2.2.10}$$

This condition is similar to (2.2.7) since it is a product of both small ($\epsilon\alpha \ll 1$) and large ($\nu/\omega_0 \gg 1$) parameters. The form (2.2.10) of the overlapping criterion can be used for the motion near the separatrix. Upon approaching the separatrix, $\alpha$ grows and $\omega_0$ vanishes. This can be found in a general form as well as for any particular case explicitly. Due to the properties of $\alpha$ and $\omega_0$, parameter $K_c$ grows and reaches the value $\sim 1$, which is somewhere close to the separatrix. The corresponding value of action $I_c$ or energy $E_c$ gives the stochastic layer a border. These qualitative comments on the separatrix destruction exclude many properties of the dynamics which will be considered in the next section using a more fundamental approach.[2]

## 2.3 The Separatrix Map

To be more specific, we continue our consideration of the perturbed pendulum model (1.4.1) which has the Hamiltonian

[2]The overlapping criterion was used in the study of chaos occurring near the separatrix for the problem of destruction of the magnetic surfaces in toroidal plasma confinement systems [FSZ 67]. The initial estimations were very cumbersome. They were improved and made universal in [ZF 68]. It was also shown that the result for the width of the stochastic layer obtained in the order of magnitude by the overlapping criterion coincides with that obtained from the separatrix map.

$$
\begin{aligned}
H &= H_0(p,x) + \epsilon V(x,t) \\
H_0 &= \frac{1}{2}p^2 - \omega_0^2 \cos x \\
V &= V_0 \cos(kx - \nu t)
\end{aligned}
\tag{2.3.1}
$$

where we put the amplitude of perturbation as $V_0$ instead of $\omega_0^2$ when comparing it with (1.4.1). For the time derivative of the pendulum's energy

$$
\dot{E} = \dot{H}_0(p,x) = [H_0, H] = \epsilon[H_0, V] = -\epsilon \frac{\partial H_0}{\partial p}\frac{\partial V}{\partial x} = -\epsilon p \frac{\partial V}{\partial x}, \tag{2.3.2}
$$

we use a notation for the Poissonian brackets:

$$
[A, B] \equiv \frac{\partial A}{\partial x}\frac{\partial B}{\partial p} - \frac{\partial A}{\partial p}\frac{\partial B}{\partial x}. \tag{2.3.3}
$$

Expression (2.3.2) is exact and general. For the pendulum with a unit mass $p = \dot{x}$, we derive either

$$
\dot{E} = \epsilon k V_0 \dot{x} \sin(kx - \nu t) \tag{2.3.4}
$$

or

$$
\Delta E(t', t'') = \epsilon k V_0 \int_{t'}^{t''} \dot{x} \sin(kx - \nu t) dt \tag{2.3.5}
$$

from (2.3.1) and (2.3.2).

In the following discussion, we use two variables: the energy $E$ and action $I$ of the unperturbed pendulum introduced in (1.4.7). The corresponding formulas are found in Appendix 1. Hence we are able to obtain $I = I(H_0)$ or its inverse relation,

$$
E = H_0 = H_0(I). \tag{2.3.6}
$$

On the separatrix,

$$
E_s = \omega_0^2 = H_s; \quad I_s = \frac{1}{\pi} 8\omega_0 E\left(\frac{\pi}{2}; 1\right) = \frac{1}{\pi} 8\omega_0. \tag{2.3.7}
$$

The separatrix map can be derived in an explicit form in the vicinity of the separatrix, that is, for

$$|E - E_s| \ll E_s\,, \quad |I - I_s| \ll I_s\,. \tag{2.3.8}$$

The idea of introducing the separatrix map is based on some general property of motion near the separatrix. The corresponding behaviour of the velocity $\dot{x} = \dot{x}(t)$ near the pendulum separatrix is presented in Fig. 1.4.3 for oscillating motion. It resembles a periodic sequence of localised pulses (solitons) because the trajectory passes near the saddle point which is stationary, that is, $\dot{x} = 0$. An important property of the frequency behaviour originates from an analysis of the motion near the separatrix for a fairly general case. The unperturbed motion satisfies the equations

$$\dot{I} = 0\,, \quad \dot{\vartheta} = \omega(I) = \omega(E) \tag{2.3.9}$$

and from (1.4.9),

$$\omega(E) = dH_0/dI = dE/dI\,. \tag{2.3.10}$$

On the separatrix with $E \equiv E_s$ and $H_0 \equiv H_s = E_s$, we have

$$\omega(E_s) = 0 \tag{2.3.11}$$

as a general condition for fairly smooth potentials. A trajectory with

$$h \equiv E/E_s - 1\,, \quad |h| \ll 1 \tag{2.3.12}$$

is very close to the separatrix and its period,

$$T(E) = 2\pi/\omega(E) = \oint \frac{dx}{[2(E - V(x))]^{1/2}}\,, \tag{2.3.13}$$

tends to infinity as

$$T(E) \propto \ln(\mathrm{const}/h)\,. \tag{2.3.14}$$

The last expression is easy to obtain if we expand the potential near the saddle point $x = x_s$:

$$V(x) = V(x_s) - \gamma(x - x_s)^2 = h - \gamma(x - x_s)^2$$

where $\gamma$ is a constant and we make use of condition (2.3.12) (see also the exact results for the pendulum (A.1.9), (A.1.13)). As we approach the separatrix, $h \to 0$ and the distance between the pulses in Fig. 1.4.3 tends to infinity while the shape of the pulses remains almost unchanged.

The structure of the unperturbed motion described permits the introduction of a separatrix map, $\hat{T}$, in the form of

$$(I_{n+1}, \vartheta_{n+1}) = \hat{T}(I_n, \vartheta_n)\,, \tag{2.3.15}$$

where the time instants of the set $\{t_j\}$ are taken at intervals $T(I)$ (or $T(I)/2$, depending on the level of convenience):

$$\{t_j\} = \{t_0, t_0 + T(I_0), \ldots\}\,. \tag{2.3.16}$$

For the unperturbed motion (2.3.9) and definition (2.3.16), the map (2.3.15) can be easily rewritten, using a time variable instead of $\vartheta$, as

$$I_{n+1} = I_n\,, \quad t_{n+1} = t_n + 2\pi/\omega(I_0) \tag{2.3.17}$$

since $I = \text{const} = I_0$. Some non-trivial steps should be made for the perturbed case (2.3.1).

First, it is necessary to mention that the time sequence (2.3.16) taken to construct the separatrix map cannot be equidistant since $I, E \neq$ const. Using energy-time variables and a half-period map, it is possible to write the map as

$$\begin{aligned} E_{n+1} &= E_n + \Delta E_n(\tau_n) \\ \tau_{n+1} &= \tau_n + \pi/\omega(E_{n+1})\,, \end{aligned} \tag{2.3.18}$$

where $\Delta E$ is the change in the energy during the time-step of the map

$$\Delta E(\tau_n) = -\epsilon \int_{\tau_n - T(E_n)/4}^{\tau_n + T(E_n)/4} dt\, \dot{x} \frac{\partial V}{\partial x} \tag{2.3.19}$$

and the sequence of time instants, $\{\tau_j\}$, corresponds to the max $|\dot{x}|$ or centres of pulses in Fig. 1.4.3. The expressions (2.3.18) and (2.3.19) are still exact. Moreover, they are written in such a way that it enables different approximations to be performed in a simple manner.

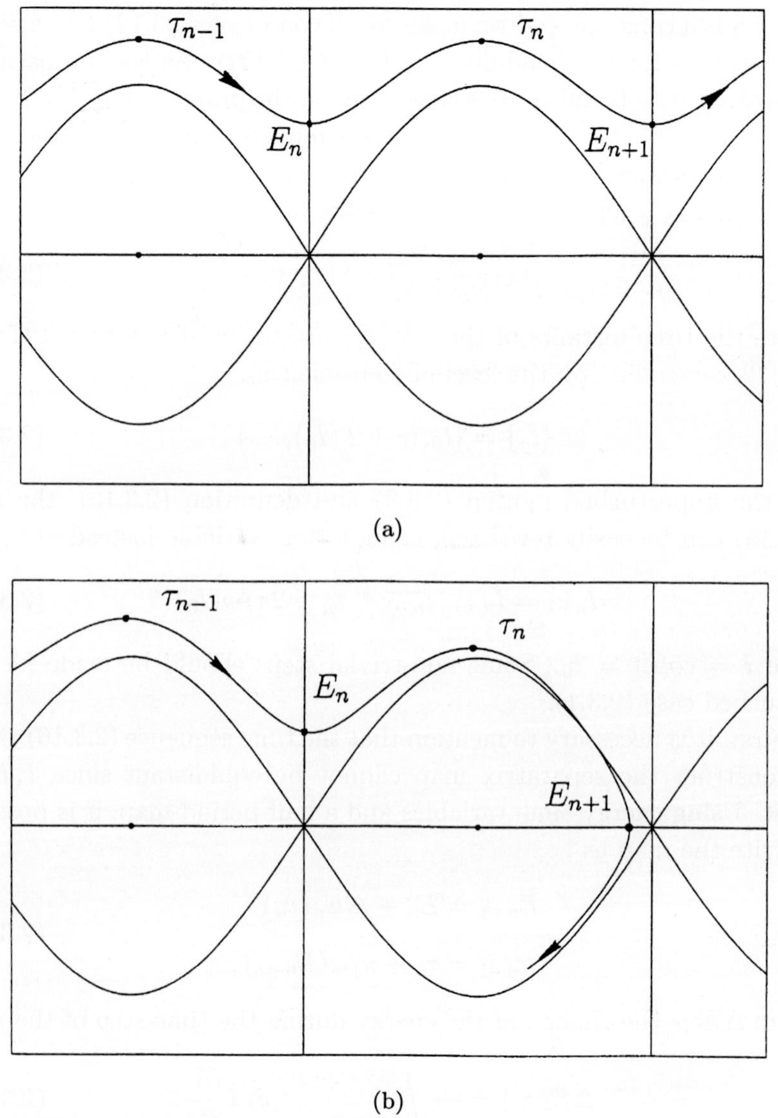


Fig. 2.3.1. Sampling points of the separatrix map for two different cases.

Taking into account the fact that the integrand expression in (2.3.19) has $\dot{x}$ with behaviour, as presented in Fig. 1.4.3, one can extend the integration in (2.3.19) to $\pm\infty$ and write $\Delta E$ as

$$\Delta E(\tau_n) = -\epsilon \int_{\tau_n - \infty}^{\tau_n + \infty} dt\, \dot{x}_s \frac{\partial V}{\partial x_s}\,. \tag{2.3.20}$$

Expression (2.3.20) is simply a Melnikov integral where we take the integrand on the separatrix as a first approximation.[3]

The values for $E_n$ are obtained for the time instants between two consequent values: $(\tau_{n-1}, \tau_n)$. This is because $E_n$ cannot be determined at the time instant $\tau_n$. Examples of two different cases of a pendulum trajectory behaviour are shown in Fig. 2.3.1 and they demonstrate how $(E_n, \tau_n)$ and $(E_{n+1}, \tau_{n+1})$ are derived. Using the dimensionless variable (2.3.12), the separatrix map (2.3.18) is rewritten as

$$\begin{aligned} h_{n+1} &= h_n + \Delta h(\tau_n) \\ \tau_{n+1} &= \tau_n + \pi/\omega(h_{n+1})\,, \end{aligned} \tag{2.3.21}$$

where

$$\Delta h(\tau_n) = -\frac{\epsilon}{E_s} \int_{\tau_n - \infty}^{\tau_n + \infty} dt\, \dot{x}_s \frac{\partial V}{\partial x_s} \tag{2.3.22}$$

is the dimensionless Melnikov integral. The separatrix map in (2.1.21) is area-preserving and the variables $(h, \tau)$ represent a canonically conjugated pair.[4]

## 2.4 Stochastic Layer

In order to understand the behaviour of $\Delta h(\tau_n)$, we consider the co-ordinate variable $x$, such as $\Delta h(x', x'')$ where the value of $x'$ corresponds to the particle position at time $t'$ and $x''$, to $t''$. For an unperturbed separatrix connecting the two saddle points (see dash curve in Fig. 2.4.1), $x'$ is considered the co-ordinate on the left saddle point and arbitrary $x'' > x'$. $\Delta h = \Delta h(x', x'')$, which is considered a function of $x''$, then surpasses zero(es) due to the oscillating properties of $V(x, t)$. Thus, the

[3] The original work is described in [Me 63] and a more simplified description is found in [ZSUC 91].

[4] The separatrix map was introduced in [ZF 68]. Later, it was given the name "whisker map" [C 79]. The different physical applications of the separatrix map are described in [ZSUC 91]. A more rigorous approach to the separatrix map is found in [Ro 94, Ro 95, Tr 97].

perturbed separatrix oscillates as shown in Fig. 2.4.1. The infinity of the period on the unperturbed separatrix induces an infinite number of oscillations as $x''$ approaches the second saddle point. The same behaviour is observed for a new saddle point: it integrates in the opposite direction. This phenomenon is known as the splitting of the separatrix. The intersections of different kinds of perturbed separatrices will create new saddle points in the first approximation. For a neighbouring pair of new saddles, the same consideration can be applied but at a higher approximation. A more complex picture will therefore emerge which points the way towards chaotic dynamics.

In considering the separatrix map, one can deduce that the two variables of $h$ and $\tau$ are not equivalent. Furthermore, while $\max \Delta h$ is limited, a change in $\tau_n$ can be arbitrarily large since $\omega(h) \to 0$ and $\omega' \to \infty$. A more simple way to derive the area of chaotic dynamics is to consider the heuristic condition

$$K = \max \left| \frac{d\tau_{n+1}}{d\tau_n} - 1 \right| \gtrsim 1 \, , \tag{2.4.1}$$

which indicates the exponential separation of trajectories. Applying (2.4.1) to (2.3.21), we obtain

$$\max \frac{\pi}{\omega^2(h)} \left| \frac{d\omega(h)}{dh} \frac{d\Delta h}{d\tau_n} \right| \gtrsim 1 \, . \tag{2.4.2}$$

It is easy to prove that (2.4.2) can always be satisfied by using a fairly small $h$. This is because $h \to 0$ means approaching the separatrix (see (2.3.12)) and $|d\omega/dh| \to \infty$ as $\omega(E)$ has a singularity at $E = E_s$. The condition

$$K = K(h_{s\ell}) = 1 \tag{2.4.3}$$

can be considered as an equation which defines the stochastic layer width, $h_{s\ell}$, up to a constant of order one. It is useful to note that the width of the stochastic layer can be obtained from a simulation using very high precision. This is because it is possible to derive the last invariant (stable) curve at an arbitrarily selected level of accuracy. In other words, the width of the stochastic layer is the distance between the last two invariant curves below and above the layer.

As an illustration of the estimation of the width of the stochastic layer, one can consider the Hamiltonian of a perturbed pendulum in the form of (2.3.1):

$$H = \frac{1}{2}\dot{x}^2 - \omega_0^2 \cos x + \epsilon\omega_0^2 \cos(x - \nu t) . \tag{2.4.4}$$

Here the mass is equal to one and the parameter $k = 1$. Hamiltonian (2.4.4) defines the corresponding equation of motion

$$\ddot{x} + \omega_0^2 \sin x = \epsilon\omega_0^2 \sin(x - \nu t) , \tag{2.4.5}$$

which occurs in many physical problems as mentioned in Section 1.4. The convenience of the separatrix map (2.3.21) lies in the fact that we need only know the motion near the separatrix. From (2.4.4) and (2.4.5), we derive

$$\begin{aligned} E_s &= \omega_0^2 \\ \dot{x}_s &= \pm 2\omega_0 / \cosh \omega_0 (t - t_n) \\ x_s &= 4 \arctan \exp\{\pm\omega_0 (t - t_n)\} - \pi \end{aligned} \tag{2.4.6}$$

on the separatrix, where the time instant $t_n$ is introduced as an initial condition (compare this to (1.4.6)). Near the separatrix, one obtains

$$\omega(E) = \omega(h) = \pi\omega_0 / \ln \frac{32}{|h|} . \tag{2.4.7}$$

This information is sufficient to calculate the Melnikov integral, $\Delta h$, in (2.3.22) and in map (2.3.21). After substituting $V = \cos(x_s - \nu(t - \tau_n))$ in (2.3.22) and omitting the fairly simple calculations, it is found that

$$\begin{aligned} \Delta h &= \epsilon M_n \sin \phi_n , \\ M_n &= 4\pi \left(\frac{\nu}{\omega_0}\right)^2 \exp\left(\sigma_n \frac{\pi\nu}{2\omega_0}\right) \Big/ \sinh \frac{\pi\nu}{\omega_0} , \\ \phi_n &\equiv \nu\tau_n (\mathrm{mod}\, 2\pi) \end{aligned} \tag{2.4.8}$$

with sign-function $\sigma_n = \pm 1$ and

$$\sigma_{n+1} = \sigma_n \cdot \mathrm{sign}\, h_n . \tag{2.4.9}$$

Using (2.4.8), the separatrix map for the perturbed pendulum is presented in the following form:

$$\begin{aligned} h_{n+1} &= h_n + \epsilon M_n \sin \phi_n \\ \phi_{n+1} &= \phi_n + \frac{\nu}{\omega_0} \ln \frac{32}{|h_{n+1}|}, \quad (\mathrm{mod}\, 2\pi). \end{aligned} \tag{2.4.10}$$

The expression for $M_n$ can be simplifed in the case of $\nu \gg \omega_0$:

$$\begin{aligned} M_n &= 8\pi \left(\frac{\nu}{\omega_0}\right)^2 \exp\left(-\frac{\pi\nu}{2\omega_0}\right), \quad \sigma_n > 0 \\ M_n &= 8\pi \left(\frac{\nu}{\omega_0}\right)^2 \exp\left(-\frac{3\pi\nu}{2\omega_0}\right), \quad \sigma_n < 0. \end{aligned} \tag{2.4.11}$$

When $\sigma_n < 0$, the value $M_n$ is much smaller than that for $\sigma_n > 0$ and it can therefore be ignored.

Finally, map (2.4.10), together with definitions (2.4.8) and (2.4.9), forms a separatrix map when $\nu$ is not too small when compared to $\omega_0$.

Combining expressions (2.4.2), (2.4.3), (2.4.7) and (2.4.8), the half-width of the stochastic layer is finally obtained:

$$h_{s\ell} = 8\pi\epsilon \left(\frac{\nu}{\omega_0}\right)^3 \exp\left(-\frac{\pi}{2}\frac{\nu}{\omega_0}\right). \tag{2.4.12}$$

The result shows that the width is exponentially small relative to the large parameter $\nu/\omega_0$. The general expression of (2.4.8) for the Melnikov integeral $M_n$ is also valid for $\nu \sim \omega_0$. In this case, the same method of estimation gives rise to

$$h_{s\ell} \sim \epsilon, \tag{2.4.13}$$

which seems to be the largest width of the stochastic layer.[5]

[5]Both estimations, the exponentially small (2.4.12) and of order $\epsilon$ (2.4.13), were obtained in [ZF 68]. They were again derived in a number of subsequent publications: [C 79], [Za 85], [LST 89] and [Tr 96]. In the latter two references, more power methods were used [see also (Tr 97)] which improved the means of obtaining the pre-exponential numerical constant in (2.4.12). A more sophisticated case is that of the adiabatic perturbation $\nu \gg \omega_0$. Special methods were developed for it: [N 75], [CET 86], [EE 91] and [NST 97].

To conclude this section, two comments need to be made:

(i) A useful characteristic of the width of the stochastic layer is

$$\delta_h = h_{s\ell}/\max \Delta h = h_{s\ell}/\epsilon M = \nu/\omega_0\,, \tag{2.4.14}$$

which is the ratio of its width to the value of the split in the separatrix (see Fig. 2.4.1). Looking at (2.4.14), it is clear that $\delta_h \gg 1$ for high frequency perturbation.

(ii) A real parameter of expansion in obtaining the value of $h_{s\ell}$ is the parameter of perturbation $\epsilon$ if $\nu \sim \omega_0$. When $\nu \gg \omega_0$, the value of $h_{s\ell}$ is exponentially small (cf. (2.4.12)) and the corrections beside (2.4.12) are of the order

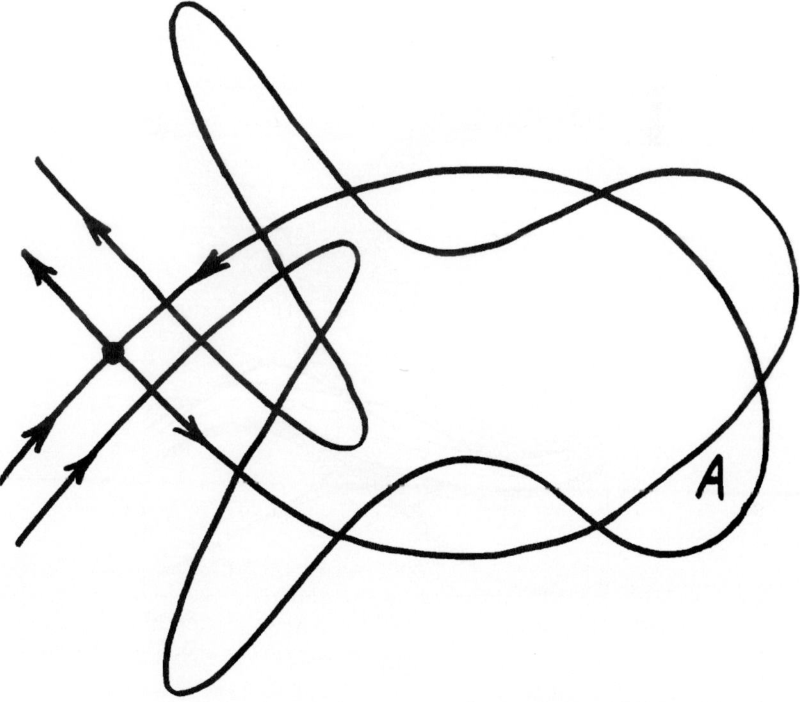


Fig. 2.4.1. Splitting of the separatrix and a lob (A).

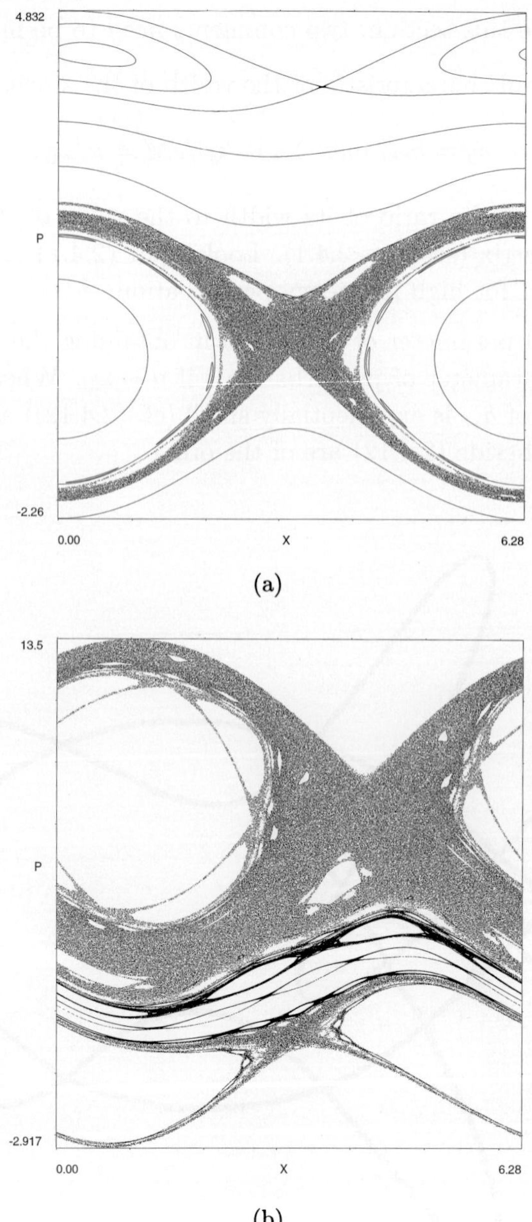


Fig. 2.4.2. Three different cases in the formation of stochastic layers: (a) small $\epsilon = 0.1$ and moderate $\nu = 4.0$; (b) $\epsilon = 8.1$; $\nu = 8.59$; and (c) $\epsilon = 5.1$; $\nu = 10.4$.

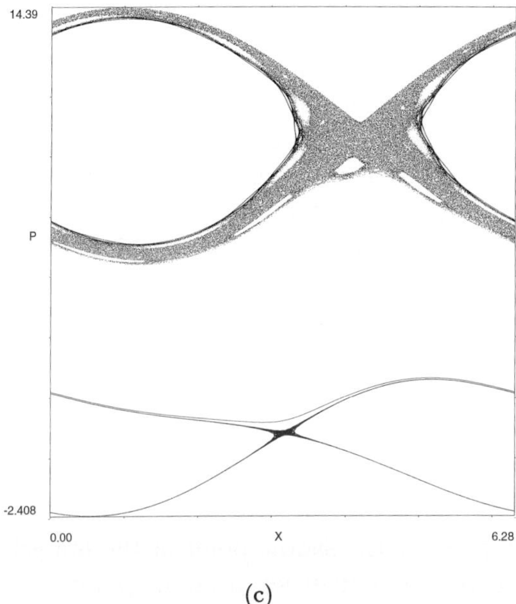


(c)

Fig. 2.4.2. (*Continued*)

$$\bar{\epsilon} = \epsilon \left( \frac{\omega_0}{\nu} \right)^2 \ll 1 \,, \qquad (2.4.15)$$

which is due to high frequency perturbation. Figures 2.4.2(a), (b) and (c) illustrate the operation of the concept of stochastic layer for small $\epsilon$ and for large $\epsilon$ with a small $\bar{\epsilon}$.[6]

## 2.5 Hidden Renormalisation Group Near the Separatrix

It would be too simplistic to remark that the perturbed motion near the separatrix is so complicated that we have yet to be able to create a more or less complete picture of the basic properties found in the stochastic

[6]Corrections of order $(\omega_0/\nu)^2$ are of a general type which occur in high frequency perturbations [LL 76]. A consideration of the width of the stochastic layer is found in [Za 94].

layer and its vicinity. This section describes the existence of a "hidden" (non-evident from the Hamiltonian) renormalisation transform. The occurrence of the transform, denoted as $R_\epsilon$, follows directly from the sepecific form of the separatrix map (2.4.10). Before the transform $R_\epsilon$ is derived, a qualitative analysis needs to be performed. It can be seen that the existence of $R_\epsilon$ is a general property of the motion near the separatrix, and therefore it is as common as the separatrix map (2.4.10).[7]

Again, a periodically perturbed system,

$$H = H_0(p, x) + \epsilon V(x, t) , \tag{2.5.1}$$

with

$$V(x, t + T_\nu) = V(x, t) , \quad T_\nu = 2\pi/\nu \tag{2.5.2}$$

is considered.

When $x = 0$, $p = 0$ is the saddle point of the unperturbed Hamiltonian $H_0(p, x)$, there is an expansion near the point

$$H_0(p, x) = \frac{1}{2}(p^2 - \omega_s^2 x^2) + \text{const} \tag{2.5.3}$$

with a constant $\omega_s$ which defines the increment in instability. The dependence, $x = x(t)$, for an unperturbed trajectory near the saddle point can be obtained through the direct integration of the equations of motion which follow from (2.5.3):

$$\omega_s t = \omega_s \int^t \frac{dx}{\dot{x}} = \omega_s \int^t \frac{dx}{p} = \omega_s \int^t \frac{dx}{(2h + x^2)^{1/2}}$$

$$= \ln |x + (2h + x^2)^{1/2}| + \text{const} \tag{2.5.4}$$

where, as in the previous section,

$$h = (H_0 - \omega_s^2)/\omega_s^2 \tag{2.5.5}$$

is the dimensionless energy derived from the separatrix value $H_s = \omega_s^2$.

[7]A hidden renormalisation transform was found for the separatrix map which describes a time-perturbed nonlinear system having a separatrix [ZA 95]. This finding was generalised to a system with two coupled nonlinear oscillators [KZ 97] on the basis of the same property of motion found near the separatrix. In this section, the data from [ZA 94] were used.

We now turn to a consideration of the renormalisation of $h$ and $x$ in (2.5.4):

$$h \to \lambda h_\lambda\,, \quad x \to \lambda^{1/2} x_\lambda\,, \quad \lambda > 0\,, \tag{2.5.6}$$

The substitution of (2.5.6) in (2.5.4) leads to the time shift

$$\Delta t = \frac{1}{2\omega_s} \ln \lambda\,. \tag{2.5.7}$$

One can therefore conclude that the existence of a periodic dependence on phase (2.5.4) will lead to the preservation of the equation of motion near the separatrix if the choice of $\Delta t$ is made in (2.5.7). Consequently, the phase space topology is preserved. This property is rigorously formulated and proven below. At this point, some general remarks are helpful:

(i) The renormalisation transform $R_\epsilon$ derived below demonstrates the existence of scaling invariance with respect to the appropriate renormalisation of the perturbation parameter $\epsilon$, that is, the renormalisation constant $\lambda$ depends on $\epsilon$.

(ii) The invariance under the transform $R_\epsilon$ appears as an approximate property of the topology for part of the phase space in the vicinity of a saddle point.

(iii) The $R_\epsilon$-transform can be applied to that part of the phase space which includes the stochastic layer and a set of resonances lying outside.

Rewritten in a more general way than (2.5.3), the expansion of $H_0(p, x)$ near a saddle point is

$$H_0 - E_s = \frac{1}{2}(A_p^2 p^2 - A_x^2 x^2) \equiv \omega_s a^+ a^-\,, \tag{2.5.8}$$

where $E_s$ is the energy at the saddle point and $A_p, A_x$ are constants. As a result, the new amplitudes,

$$a^\pm = (2\omega_s)^{-1/2}(A_p p \pm A_x x)\,, \tag{2.5.9}$$

and an imaginary "frequency",

$$\omega_s = A_p A_x\,, \tag{2.5.10}$$

are introduced. Hamiltonian (2.5.8) describes the hyperbolic motion with increment (decrement) $\omega_s$:

$$\dot{a}^{\pm} = \pm\frac{\partial H_0}{\partial a^{\mp}} = \pm\omega_s a^{\pm}\,, \tag{2.5.11}$$

that is,

$$a^{\pm}(t) = a^{\pm}(0)\exp(\pm\omega_s t)\,. \tag{2.5.12}$$

The saddle point (0,0) belongs to a singular curve, the separatrix. Only the motion on the surface of a cylinder phase space is considered. This means that two different periodic motions are separated by the separatrix. For fairly smooth Hamiltonians, the nonlinear frequency near the separatrix can be written in a way that is similar to (2.4.7):

$$\omega(h) = A\omega_s/\ln(B/|h|) \tag{2.5.13}$$

where $A$ and $B$ are some constants depending on the form of $H_0(p,x)$ and

$$h = (E - E_s)/E_s\,, \tag{2.5.14}$$

which is similar to (2.5.5) with energy $E = H_0$. In the example of (2.4.7), $\omega_s = \omega_0$, $A = \pi$ and $B = 32$.

The separatrix map (2.4.10) can be rewritten in a more general form as

$$\begin{aligned} h_{n+1} &= h_n + \epsilon M(n,\phi_n) \\ \phi_{n+1} &= \phi_n + \frac{\pi\nu}{\omega_s A}\ln\frac{B}{|h_{n+1}|}\,, \quad (\mathrm{mod}\,2\pi) \end{aligned} \tag{2.5.15}$$

where $\epsilon M(n,\phi_n)$ is the Melnikov integral $\Delta h(\phi_n)$ in (2.3.22), $\phi_n = \nu\tau_n$, and expression (2.5.13) has been used.

**Proposition (1S):** The separatrix map (2.5.15) is invariant under the transform

$$R_\epsilon\colon \epsilon \to \lambda\epsilon\,, \quad h \to \lambda h \tag{2.5.16}$$

if

$$\lambda = \exp(2\omega_s A/\nu)\,. \tag{2.5.17}$$

It also follows from (2.2.16) that

$$R_\epsilon^m\colon \epsilon \to \lambda^m \epsilon\,, \quad h \to \lambda^m h\,, \quad m = \pm 1, \pm 2, \ldots \tag{2.5.18}$$

with the same $\lambda$ (2.5.17).

The statement (1S) follows from (2.5.15) to (2.5.17) if one takes into account the fact that the phase $\phi_{n+1}$ is defined in (2.5.15) up to the term $2\pi m$, where $m$ is an integer. In other words, the renormalisation transform $R_\epsilon$ (2.5.16) preserves the phase space topology near the saddle point with the same level of accuracy with which the separatrix map (2.5.15) was obtained.

In order to formulate an invariancy that is similar to (1S) for the original Hamiltonian (2.5.1), one can recall that in the construction of the separatrix map (see Section 2.3) the time instants $\tau_n$ correspond to the midway point between the two adjacent passages of a saddle point, say $\tau_n^{(s)}$ and $\tau_{n+1}^{(s)}$. This means that

$$\tau_n = \tau_n^{(s)} + \pi/2\omega(h_n) \tag{2.5.19}$$

or, after multiplying by $\nu$,

$$\phi_n = \phi_n^{(s)} + \pi\nu/2\omega(h_n)\,, \quad \phi_n^{(s)} = \nu\tau_n^{(s)}\,. \tag{2.5.20}$$

We also need to recall that the magnitude $h_n$ is taken at precisely the time instant $\tau_n^{(s)}$ of the passage near the corresponding saddle point, that is,

$$h_n = h(\tau_n^{(s)}) \tag{2.5.21}$$

by definition. The substitution of (2.5.20) in (2.5.15) gives rise to

$$\begin{aligned} h_{n+1} &= h_n + \epsilon M(n, \phi_n^{(s)} + \pi\nu/2\omega(h_n)) \\ \phi_{n+1}^{(s)} &= \phi_n^{(s)} + \frac{\pi\nu}{\omega_s A} \ln \frac{B}{|h_{n+1}|} + \frac{\pi\nu}{2\omega_s A} \ln \left| \frac{h_{n+1}}{h_n} \right|, \quad (\operatorname{mod} 2\pi)\,. \end{aligned} \tag{2.5.22}$$

Map (2.5.22) is called the shifted separatrix map (SSM). Its specific property is that both variables $(h, \phi^{(s)})$ are taken at the same time instant.

One is therefore ready to formulate the invariant properties of the SSM.

**Proposition (2S):** The SSM (2.5.22) is invariant under the renormalisation transform

$$R_\epsilon\colon \epsilon \to \lambda\epsilon\,, \quad h \to \lambda h\,, \quad \phi^{(s)} \to \phi^{(s)} + \pi \tag{2.5.23}$$

if

$$\lambda = \exp(2\omega_s A/\nu)\,. \tag{2.5.24}$$

The difference between (2S) and (1S) is that there is a phase shift of $\phi^{(s)}$ by $\pi$ in the case of (2.5.23) where both variables $(h, \phi^{(s)})$ are taken at the same time instant. The results, (2.5.23) and (2.5.24), can be applied directly to the Hamiltonian (2.5.1).

**Proposition (3S):** The phase portrait topology near the saddle point, obtained as a Poincaré map of the Hamiltonian (2.5.1), preserves

$$\begin{gathered} R_\epsilon\colon \epsilon \to \lambda\epsilon\,, \quad H_0 \to \lambda H_0\,, \quad t \to t + \pi/\nu \\ x \to \lambda^{1/2}x\,, \quad p \to \lambda^{1/2}p \\ \lambda = \exp(2\omega_s A/\nu) \end{gathered} \tag{2.5.25}$$

under the renormalisation transform. The proof for (3S) is based on the expansion (2.5.8) for $H_0$ near the saddle point.

It follows from (3S) that the scaling transform (3S′)

$$\begin{gathered} R_\epsilon^m\colon \epsilon \to \lambda^m\epsilon\,, \quad H_0 \to \lambda^m H_0\,, \quad t \to t + m\pi/\nu \\ x \to \lambda^{m/2}x\,, \quad p \to \lambda^{m/2}p\,, \quad m = \pm 1, \pm 2 \end{gathered} \tag{2.5.26}$$

preserves the phase portrait near the saddle point with the same $\lambda$ (2.5.25).

(3S′) is analogous to (1S′) and is valid up to $m < m_0$ when the accumulation of errors becomes significant.

Figure 2.5.1 illustrates the properties (1S) and (3S) for the perturbed pendulum model (2.4.4). In this model, $\omega_s = \omega_0$, $A = \pi$ and

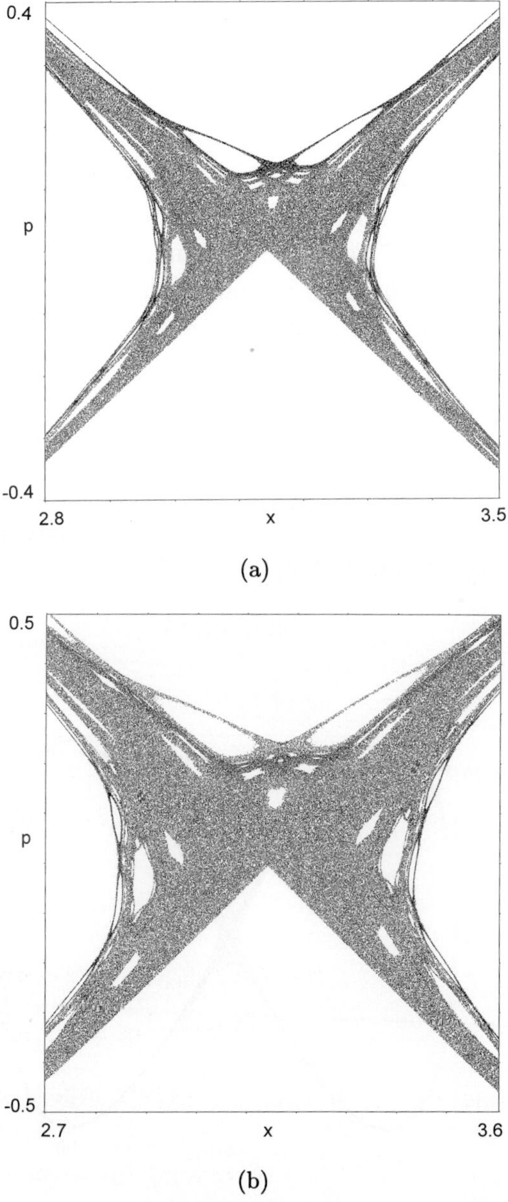


Fig. 2.5.1. Poincaré section of motion in the ergodic layer: (a) before a renormalisation ($\nu = 5.4$; $\omega_0 = 1$; $\epsilon = 0.01$); and (b) after the renormalisation ($\epsilon = 0.032103$; $\lambda = 3.2013$).

$$\lambda = \exp(2\pi\omega_0/\nu)\,. \tag{2.5.27}$$

In the numerical example given in Fig. 2.5.1, we have $\lambda = 3.2013$. Hence the difference between the values of perturbation constant $\epsilon$ and $\epsilon/\lambda^2$ for cases (a) and (b) is almost of one order.

There are two remarkable properties in the renormalisation transform $R_\epsilon$: it is not very obvious in the Hamiltonian (and that is why we use the word "hidden") and the renormalisation parameter $\lambda$ can be obtained directly from the Hamiltonian of the types (2.5.1) and (2.5.2). As another example, for a Hamiltonian with a double-well potential

$$\begin{aligned} H_0 &= \frac{1}{2}p^2 - \frac{1}{2}\gamma^2 x^2\left(1 - \frac{x^2}{2}\right) \\ \epsilon V &= \frac{1}{4}\epsilon x^4 \sin \nu t\,, \end{aligned} \tag{2.5.28}$$

$\lambda$ is of the same type as in (2.5.27):

$$\lambda = \exp(2\pi\gamma/\nu)\,. \tag{2.5.29}$$

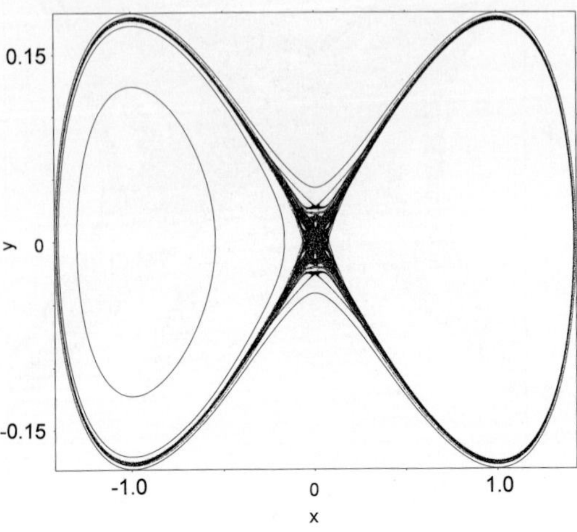


Fig. 2.5.2. Stochastic layer for a double-well potential: $\epsilon = 8.3 \cdot 10^{-5}$, $a = 1$, $\alpha = 0.25$ and $\beta = 1$.

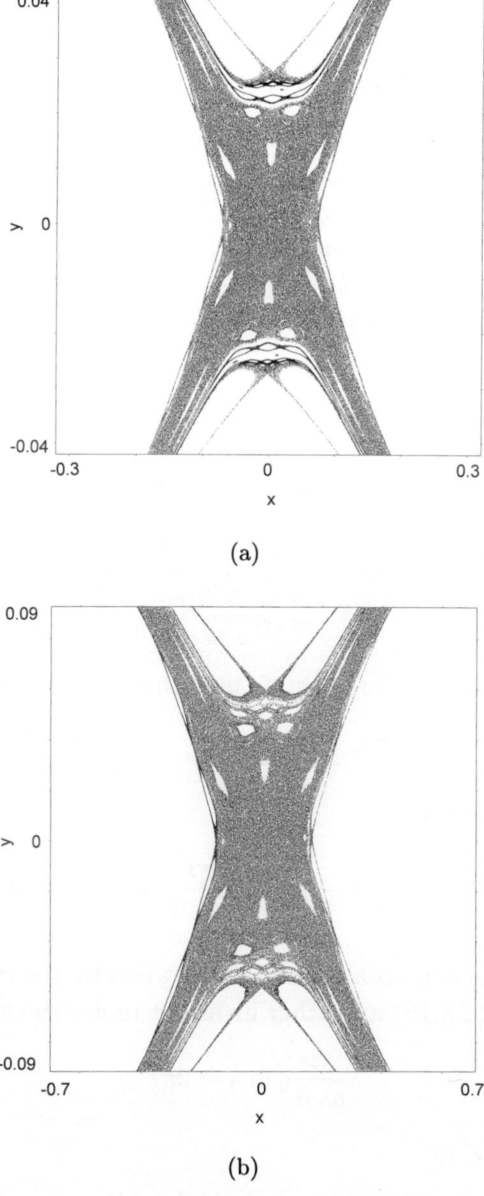


Fig. 2.5.3. The vicinity of the saddle point for the same model as presented in Fig. 2.5.2: (a) before the renormalisation (see parameters in Fig. 2.5.2); and (b) after the renormalisation with $\lambda_\epsilon = 4.81$ and the phase shifted by $\pi$.

Furthermore, for the numerical example in Fig. 2.5.2, we have $\gamma/\nu = 0.25$ and $\lambda = 4.8$. Figure 2.5.3 reveals a strong coincidence in the topology near the saddle point for both the original and renormalised phase portraits.

## 2.6 Renormalisation of Resonances

The transform $R_\epsilon$ can be applied to the stochastic layer as well as the resonances lying beyond it (though not too far from the separatrix).

A resonance condition can be written for the phase equation (2.5.15) as

$$\phi_{n+q} = \phi_n + 2\pi s\,, \tag{2.6.1}$$

where $s$ is an integer. By definition, Eq. (2.6.1) corresponds to the $(q, s)$-order resonance. When the phase equation of (2.5.15) is iterated $q$ times, it yields

$$\phi_{n+q} = \phi_n + \frac{\pi\nu}{\omega_s A}\sum_{j=1}^{q} \ln \frac{B}{|h_{n+j}(q,s)|}\,. \tag{2.6.2}$$

Turning now to a consideration of the formal transform

$$h = h'/\lambda\,, \tag{2.6.3}$$

we apply it to (2.6.2):

$$\phi_{n+q} = \phi_n + \frac{\pi\nu}{\omega_s A}\sum_{j=1}^{q} \ln \frac{B}{|h'_{n+j}|} + \frac{\pi\nu}{\omega_s A} q \ln\lambda\,, \quad (\text{mod}\, 2\pi)\,. \tag{2.6.4}$$

From (2.6.4), one can obtain a condition whereby the resonances of the separatrix map (2.5.15) are either invariant under the transform (2.6.3)

$$\frac{\pi\nu}{\omega_s A} q \ln\lambda = 2\pi s' \tag{2.6.5}$$

with an integer $s'$ or

$$\lambda(q, s') = \exp\left(\frac{2s'}{q}\frac{\omega_s A}{\nu}\right). \tag{2.6.6}$$

In particular, for the perturbed pendulum (2.3.1), $A = \pi$ and $\omega_s = \omega_0$. Moreover, instead of (2.6.6), we have

$$\lambda(q, s') = \exp\left(2\pi \frac{s'}{q} \frac{\omega_0}{\nu}\right). \tag{2.6.7}$$

In the case when $s' = q$, we have the same renormalisation constant as (2.5.17) and (2.5.24) as well as the transformation of the resonance family (2.6.1) into the family

$$\phi_{n+q} = \phi_q + 2\pi s'' \,, \; s'' = s + s' \,. \tag{2.6.8}$$

Condition (2.6.8) should be used together with (2.6.4) to determine the resonance energies $h' = h'(q, s'')$. The result can be formulated as follows:

**Proposition (4S):** When the conditon (2.6.1) defines a resonance family of order $(q, s)$ and Eq. (2.6.2) defines the corresponding resonance values of energy $h = h(q, s)$, it leads to the existence of a family of renormalisation constants, $\lambda = \lambda(q, s')$ (2.6.6), which transform the resonance of order $(q, s)$ into $(q, s'')$ with energy $h' = h'(q, s'')$, thus satisfying Eq. (2.6.4).

This renormalisation of resonances near the separatrix is also hidden, that is, it is not obvious from the initial Hamiltonian.

## 2.7 Stochastic Layer of the Standard Map

The standard map was introduced in Section 1.2 by the Hamiltonian (1.2.1):

$$H = \frac{1}{2} I^2 - K \cos\theta \sum_{n=-\infty}^{\infty} \delta\left(\frac{t}{T} - n\right). \tag{2.7.1}$$

We are, however, mainly interested in small perturbations:

$$K \ll 1 \,. \tag{2.7.2}$$

Nonetheless, even with this simplification, Eq. (2.7.1) still appears extremely complicated. This can be seen in Fig. 1.2.2 which demonstrates the phase portrait of system (2.7.1) for small $K$. It comprises the following elements: after the separatrix has been destroyed, a stochastic layer forms in its place. Inside the layer, a family of nested invariant curves which confine the point $[0, 0(\text{mod } 2\pi)]$ reside. Outside the main stochastic layer, chains of separatrix loops with considerably thinner stochastic layers exist which, though not visible in Fig. 1.2.2(a), can be seen after it has been magnified in Fig. 1.2.2(b). The chains correspond to nonlinear resonances of different orders and invariant curves lie between them. Thus, the phase portrait resembles a sandwich composed of an infinite number of invariant curves alternating with the stochastic layers. In other words, the stochastic layers do not merge for small values of $K$.

There is a more simple way of observing the formation of stochastic layer for the standard map. To do so, the Hamiltonian (2.7.1) is rewritten in the form (1.2.5)

$$H = \frac{1}{2}I^2 - K\cos\theta \sum_{m=-\infty}^{\infty} \cos m\nu t \tag{2.7.3}$$

and only the terms with $m = 0, \pm 1$ in (2.7.3) were kept. Thus, we obtain the following:

$$H \approx \frac{1}{2}I^2 - K\cos\theta - 2K\cos\nu t\cos\theta\,. \tag{2.7.4}$$

Equation (2.7.4) describes the pendulum's oscillations at a dimensionless frequency of

$$\omega_0 = K^{1/2}\,. \tag{2.7.5}$$

The destroyed separatrix of these oscillations forms the main stochastic layer in Fig. 1.2.2. The perturbation is given by the third term in (2.7.4). It has the same amplitude as the main term in the Hamiltonian $H$. However, its dimensionless frequency,

$$\bar{\nu} = 2\pi\,, \tag{2.7.6}$$

is large when compared to $\omega_0$, due to Eqs. (2.7.2) and (2.7.5). This enables us to make use of the formula (2.4.12) for the width of the stochastic layer using the comment which introduces the real small parameter (2.4.15). After substituting

$$\epsilon \to 2K\,, \quad \omega_0 \to K^{1/2}\,, \quad E_s = \omega_0^2 \to K\,, \quad \nu \to \bar{\nu} = 2\pi \tag{2.7.7}$$

in (2.4.12), the following is obtained:

$$h_{s\ell} = \frac{(4\pi)^4}{2K^{1/2}} \exp(-\pi^2/K^{1/2})\,. \tag{2.7.8}$$

Equation (2.7.8) defines an exponentially narrow stochastic layer for any small value of $K$. Should the subsequent terms with $m = \pm 2$ be retained in the sum in (2.7.3), the resulting corrections in Eq. (2.7.8) will be exponentially small. This is because the effective frequency of perturbation due to these terms is $\bar{\nu} = 4\pi$. Thus, the definition of the exponential term in (2.7.8) is very accurate.[8]

An important comment on the estimation (2.7.8) is that the coefficients for the main term, $K\cos\theta$, in (2.7.4) and the perturbation are of the same order, that is, the effective perturbation parameter is of the order one. Nevertheless, the minuteness of the result for the width of the stochastic layer is attributed to the exponential factor in (2.7.8).

The approximation (2.7.4) can be used to prove the absence of the standard map in the hidden renormalisation transform $R_\epsilon$ described in Section 2.5. After introducing a dimensionless parameter,

$$\tau = 2\pi/K^{1/2} > 1\,, \tag{2.7.9}$$

and rewriting the separatrix maps (2.4.8)–(2.4.10) for the Hamiltonian (2.7.4) using the replacements (2.7.7), one derives

$$\begin{aligned}
h_{n+1} &= h_n + 4\pi\tau^2(1+\sigma_n)\exp\left(-\frac{1}{2}\pi\tau\right)\sin\phi_n \\
\phi_{n+1} &= \phi_n + \tau\ln(32/|h_{n+1}|)\,, \quad (\mathrm{mod}\, 2\pi) \\
\sigma_{n+1} &= \sigma_n \cdot \mathrm{sign}\, h_n
\end{aligned} \tag{2.7.10}$$

where all smaller terms are neglected. After rescaling $\tau$: $\tau \to \lambda_\tau \tau$, that for $h$ is

$$\begin{aligned}
h &\to \lambda_h h \\
\lambda_h &= \lambda_\tau^2 \exp\left(-\frac{1}{2}\pi(\lambda_\tau - 1)\tau\right).
\end{aligned} \tag{2.7.11}$$

---

[8]The result (2.7.8) is found in [C 79].

Substituting (2.7.11) in (2.7.10), the following is obtained:

$$\phi_{n+1} = \phi_n + \lambda_\tau \tau [\ln(32/|h_{n+1}| - \ln \lambda_h] \,, \quad (\mathrm{mod}\, 2\pi) \,. \qquad (2.7.12)$$

Expression (2.7.12) shows that we cannot find such a $\lambda_\tau$ from the equation

$$(\lambda_\tau \tau - 1) \ln(32/|h_{n+1}|) - \tau \lambda_t \cdot \ln \lambda_h = 0 \,, \quad (\mathrm{mod}\, 2\pi) \,,$$

which is independent of $h$. In other words, the lack of independent parameters does not permit renormalisation.

## Conclusions

1. Perturbation changes the phase space topology of systems. Resonant perturbation creates chains of islands and increases the number of elliptic and hyperbolic points in the phase space. The overlap between resonance chains of islands creates chaotic dynamics (Chirikov criteria).
2. The perturbation of motion near a separatrix can be described using a universal separatrix map,

$$\hat{M}_s \colon \begin{array}{ll} E_{n+1} & = E_n + \Delta E_n(\tau_n) \\ \tau_{n+1} & = \tau_n + \pi/\omega(E_{n+1}) \end{array} \,,$$

   for energy $E_n$ and specially taken time instants, $\tau_n$. The specific properties of $\hat{M}_s$ are imposed by the dependence of frequency $\omega(E)$ on $E$ near the separatrix

$$\omega(E) = A / \ln \frac{B}{|E - E_s|} \,.$$

   The appropriate constants of $A$ and $B$ and the value of energy, $E_s$, define the separatrix. The perturbation of energy, $\Delta E$, is obtained from the so-called Melnikov integral.

3. The analysis of the separatrix map shows that perturbations destroy the separatrix and replace it with a stochastic layer. The width of the stochastic layer can be obtained from $\hat{M}_s$. The stochastic layer is a seed of chaos in Hamiltonian systems. It seems that $\hat{M}_s$ describes the main mechanism by which chaos originates in Hamiltonian dynamics.

4. The stochastic layer and $\hat{M}_s$ possess a renormalisation property with respect to the parameter of perturbation $\epsilon$ in the Hamiltonian

$$H = H_0(p, x) + \epsilon V(x, t) .$$

The corresponding formulation states that the phase portrait topology near the saddle point persists under the transformation

$$\hat{R}_\epsilon: \begin{array}{l} \epsilon \to \lambda\epsilon, \ H_0 \to \lambda H_0, \ t \to t + \pi/\nu \\ x \to \lambda^{1/2}x, \ p \to \lambda^{1/2}p \end{array}$$

with a specially chosen rescaling parameter,

$$\lambda = \exp(2\omega_s A/\nu) ,$$

and known parameters $\omega_s, A, \nu$ in the system. The expression for $\lambda$ is not evident from the Hamiltonian and for that reason, $\hat{R}_\epsilon$ is called the hidden renormalisation.

## *Chapter 3*

# THE PHASE SPACE OF CHAOS

## 3.1 Non-universality of the Scenario

The idealised system with chaotic dynamics has a phase space uniformly filled with any trajectory that reveals both ergodic and mixing properties independent of the initial condition in the trajectory of up to zero measure exclusions. The Arnold cat area-preserving map

$$x_{n+1} = 2x_n + y_n\,, \quad y_{n+1} = x_n + y_n\,, \quad (x, y \bmod 1) \tag{3.1.1}$$

is an example of such a system. The physical reality is so different from the case (3.1.1) that it does not permit an understanding of the many properties of chaotic trajectories. It also does not provide answers to several questions, such as:

(i) Is the measure of chaotic orbits finite (non-zero)?
(ii) How many different measures (or distribution functions) exist in the domain of chaotic motion?

The major difficulty stems from the existence an infinite number of elliptic points and islands with invariant stable curves confining the former. This phenomenon was demonstrated in Figs. 1.2.2(b), 1.3.1(b) and 1.4.1.

The presence of islands renders the chaotic motion non-ergodic. The difficulty of finding a reasonable description for chaotic dynamics also makes it difficult to gain an understanding of the importance of non-ergodicity. Initially, it was believed that after one has excluded the islands set, the remaining domain would consist of the ergodic motion with normal (in some sense) properties, such as system (3.1.1). Unfortunately, one needs to appreciate that:

(iii) The measurement of the remaining domain is not known after the island set has been extracted, and that one cannot exclude the possibility that it may be zero.
(iv) The behaviour of the trajectories around the boundaries of the islands is different from that further afield. The boundaries are "sticky" and the stickiness depends on factors such as the type of islands and parameters. Therefore, a singular zone formed by the boundary layers of the islands is said to exist.

Given the complex situation and lack of understanding of chaotic dynamics, some general features and scenarios can be introduced with a reasonable level of certainty. It seems that the splitting of the separatrix and the creation of the stochastic layer are a typical seed of chaos in Hamiltonian systems. One is therefore urged to trace the locations and connection of the separatrices in the phase space while considering the net of separatrices as a skeleton for the appearance of chaos. A common difficulty in finding a skeleton arises because in an attempt to ease the problems of doing so, one prefers to consider the small perturbations in systems close to the integrable ones. This means that sometimes unperturbed systems do not have any separatrices. Also, the separatrices sometimes occur after the effect of a perturbation has taken place and, more precisely, they reveal themselves in the destroyed form of thin stochastic layers. Examples of this phenomenon include the kicked rotator and oscillator. Both display two different scenarios: the standard map has thin layers strongly separated from each other in the $p$-direction (see Fig. 1.2.2(a)) while the web-map has a stochastic web that covers the entire phase space in both directions like a connected net (see Fig. 1.3.1(a)). In both cases, the thickness of the stochastic layers and the perturbation parameter tend to zero without any changes to their skeleton topology. However, there are instances where the skeleton topology depends significantly on $\epsilon$ when $\epsilon \to 0$.

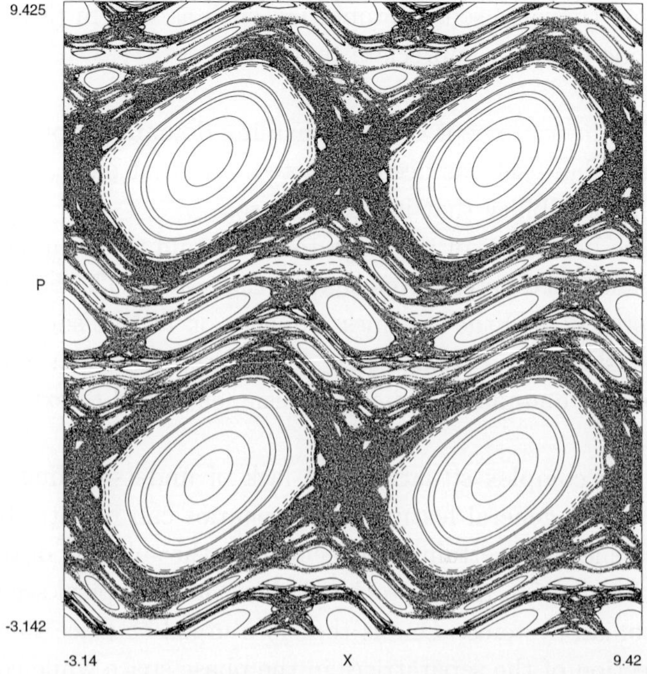


Fig. 3.1.1. Stochastic sea and islands for the standard map with overlapping resonances ($K = 1$).

Consider what happens when $\epsilon$ increases. For the standard map (1.2.8), the equivalent of the $\epsilon$ parameter is $K$, and for $K > K_c = 0.9716\ldots$, the stochastic layers merge (as in Fig. 3.1.1), thus creating a connected domain of chaotic motion that covers the entire phase plane and permits the unbounded propagation of particles in the $p$-direction.[1] When $K$ continues to increase, the visible size of the islands decreases and only the largest of them can be found in the "stochastic sea". In fact, the islands play an extremely complicated bifurcation game, such as splitting, appearing and disappearing. Some of the properties of the islands are discussed in Section 3.2.

---

[1] A theory on how to find $K_c$ was put forth by J. Green [Gr 79]. It works well at least for the standard map. Much has been written on this subject (see, for example, [MMP 84] and the reviews in [LiL 92] and [M 92]).

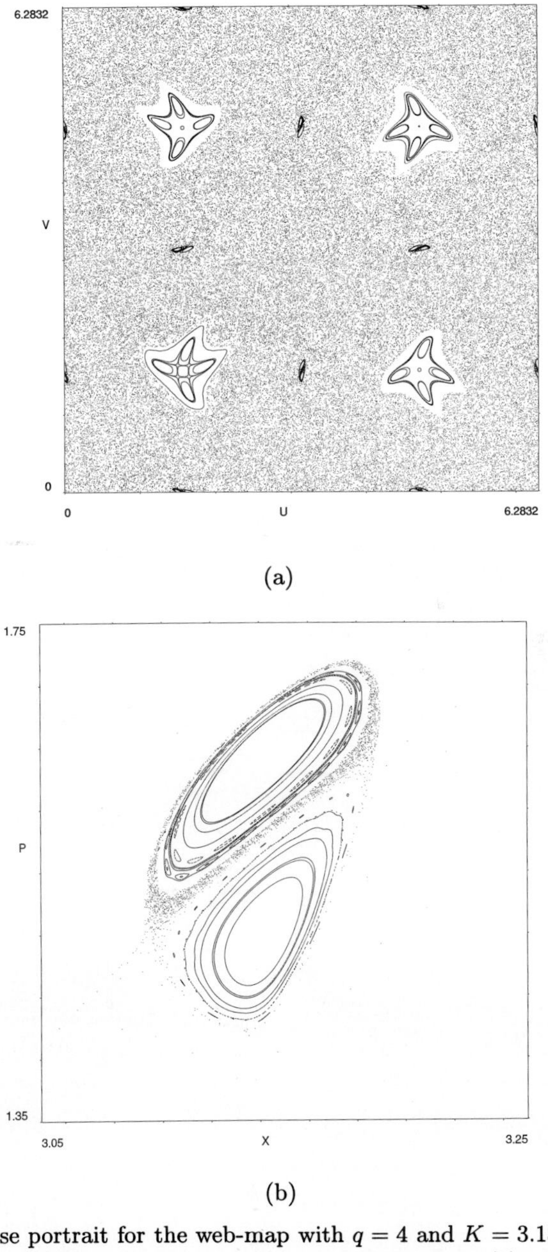

(a)

(b)

Fig. 3.1.2. Phase portrait for the web-map with $q = 4$ and $K = 3.15$: (a) stochastic sea and islands; and (b) magnification of the small islands in (a) which corresponds to an accelerator mode.

The transition of the separated stochastic layers to the stochastic sea described earlier does not exist for the web-map (1.3.5) or (1.3.12). This is because the meshes of the stochastic web cover the entire phase plane. The islands for the value of $K = 3.15$ (Fig. 3.1.2(a)) are sufficiently small and we can say that they are embedded in the stochastic sea. The distribution of the points of a trajectory in the stochastic sea looks fairly uniform if we exclude the area of the islands. Nevertheless, the boundary of the bottom left island has a dark area, indicating that it is either a more frequently visited area or, as mentioned above, it could be due to the phenomenon of the stickiness of the island. In Fig. 3.1.2(b), we present a magnified view of a tiny island found in Fig. 3.1.2(a). Its area is smaller than that of a large island in Fig. 3.1.2(a) by more

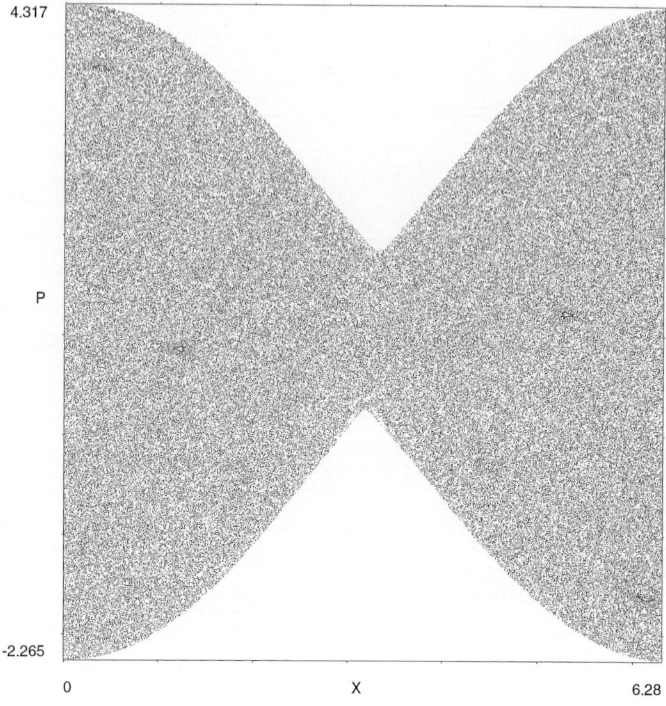


Fig. 3.1.3. Phase space of the perturbed pendulum with almost "disappeared" islands ($k = \omega_0 = 1$; $\nu = 2.07$, $\epsilon = 1$).

than two orders. At the same time, the influence of tiny islands on the process of diffusion is much stronger, which reveals how deceptive an oversimplified consideration of chaotic dynamics can be. The significant difference in the roles of the islands is due to their different properties, which are considered in Section 3.2.

Two more examples are included here. Besides increasing the variety of the situations in the phase plane, the non-trivial role of non-ergodicity in the phase space is emphasised. In the perturbed pendulum model (1.4.3) with $k = \omega_0 = 1$, a numerical simulation will reveal that for a small domain $(\delta\nu, \delta\epsilon)$ near the values $(\nu = 2, \epsilon = 1)$, there exists many points $(\nu_j, \epsilon_j) \in (\delta\nu, \delta\epsilon)$ such that the visible islands disappear. Figure 3.1.3 shows the phase portrait with very small spots where the islands are located. By adjusting the parameter $\nu$ (or $\epsilon$) finely, we can make the size of the islands arbitrarily small. All the islands make up a one-resonant system. Nevertheless, we cannot claim that the islands of a higher order resonance disappear simultaneously with the main set of islands.

A more sophisticated example is related to the Sinai billiard of the so-called infinite horizon type. A typical picture of its trajectory is given in Fig. 3.1.4. Any trajectory has an infinite number of arbitrarily long parts that correspond to a particle which bounces almost vertically (or horizontally), as shown in Fig. 3.1.4(a). In the periodically continued billiard, that is, for the corresponding Lorentz gas, the long-lived bounces correspond to long "flights" without any scattering. The Poincaré map for the billiard trajectories can be introduced in different ways. A co-ordinte, $x$, is used when the trajectory hits the lower horizontal side and the velocity $v_x$ is at the same time instant. The corresponding phase portrait in Fig. 3.1.4(b) does not have islands. The motion which is ergodic in the distribution of points in the Poincaré section seems constant. Nevertheless, one can see four scars of zero-measure which are due to the stagnation of the trajectory near the singular zones that correspond to infinite bouncing without any scattering (infinite flights). Two of the scars are related to horizontal and vertical bouncings, and two others are related to the oblique motion. In fact, an infinite number of such scars exist in the phase space of the billiard. The

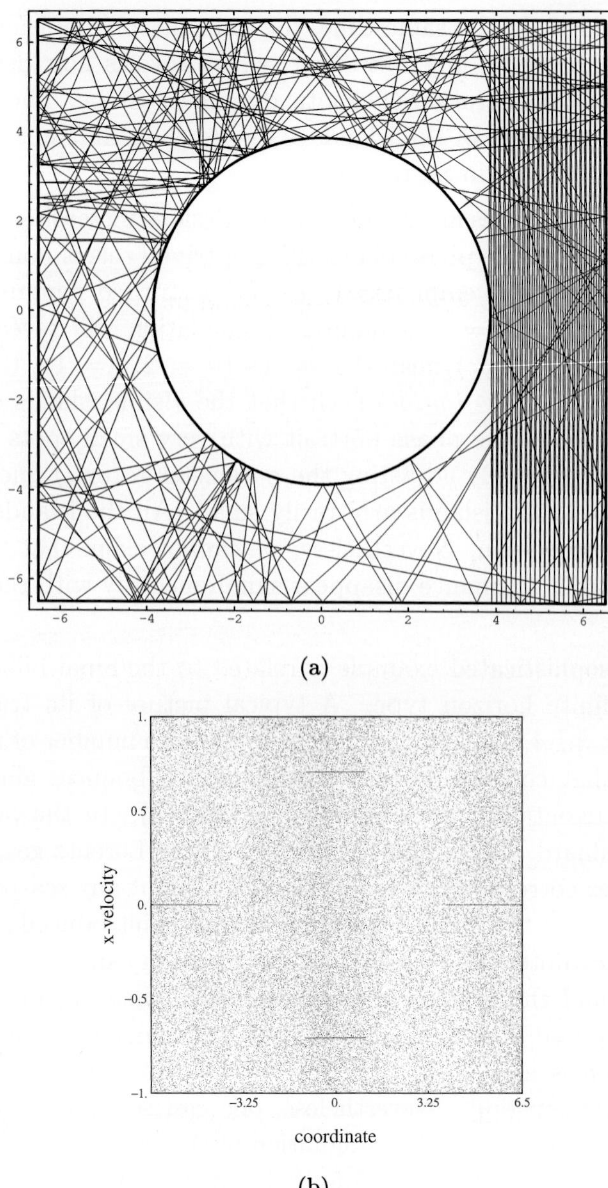


Fig. 3.1.4. Sinai billiard: (a) trajectory of the bouncing particle with a long-lasting undispersed motion; and (b) Poincaré section for the same as in (a) trjaectory with "scars" that correspond to the flights.

zero-measure domain which creates non-ergodicity strongly influences all the kinetics and long-term characteristics of the motion, as shown in Sections 7.3 and 7.4.

## 3.2 Collapsing Islands

An island with an adjoining boundary layer is called a singular zone. In fact, the dynamics in the entire phase space is influenced by an island. Nevertheless, the local structure of an island and its vicinity are sufficient (at least sometimes) in predicting large-scale asymptotics. In other words, the possibility of finding a connection between the local properties of the singular zone and large-scale properties such as kinetics and transport will be explored. This gives rise to an interest in the origin, structure and transformation of islands. At present, one is not able to classify the various types of islands. However, the few islands which are important to the study of the transport problem is described in this chapter.

Islands embedded in the area of stochastic motion correspond to some resonances between unperturbed motion and perturbations. As the perturbation parameter changes, the topology of the islands also change and different bifurcations follow one after another. One is interested in observing the birth (or death) of an island at some critical value of the perturbation parameter. One type of related islands correspond to the so-called accelerator modes.[2]

In considering the standard map (1.2.8),

$$p_{n+1} = p_n - K \sin x_n \,, \quad x_{n+1} = x_n + p_{n+1} \,, \quad (\mathrm{mod}\, 2\pi) \qquad (3.2.1)$$

its phase space can be presented as the fundamental domain ($0 \leq p \leq 2\pi$, $0 \leq x \leq 2\pi$) which is periodically continued in $x$ and $p$ directions as shown in Fig. 3.1.1. An attempt can be made to find a solution that corresponds to the ballistic motion along either $x$- and $p$- or along both the $x$- and $p$-directions. For example, when

$$p_0^{(a)} = 2\pi m \,, \quad K \sin x_n^{(a)} = 2\pi \ell \,, \quad (\ell \geq 1) \,, \qquad (3.2.2)$$

[2]For more details on the accelerator modes, see [LiL 92] and [IKH 87].

where $m$ and $\ell$ are integers, the solution

$$p_n^{(a)} = 2\pi\ell \cdot n + p_0^{(a)} \qquad (3.2.3)$$

is known as the accelerator mode. This is because $p$ grows linearly with time (number of iterations $n$) and, correspondingly, $x$ grows as $t^2$.

The values (3.2.2) and (3.2.3) can be stirred slightly without destroying the existence of the accelerator mode. To find the condition for its stability, one should consider the tangent matrix to (3.2.1)

$$M' = \begin{Vmatrix} 1 & K\cos x \\ 1 & 1 + K\cos x \end{Vmatrix} \qquad (3.2.4)$$

and find its eigenvalues $\lambda$. From the equation

$$\lambda^2 - 2\lambda\left(1 + \frac{1}{2}K\cos x\right) + 1 = 0\,, \qquad (3.2.5)$$

one derives

$$\lambda = 1 + \frac{1}{2}K\cos x \pm \left[\left(1 + \frac{1}{2}K\cos x\right)^2 - 1\right]^{1/2} \qquad (3.2.6)$$

and the stability condition is

$$0 > K\cos x > -4\,. \qquad (3.2.7)$$

Combining (3.2.7) and (3.2.2), the stability domain for the accelerator mode is obtained:

$$2\pi\ell < K < [4^2 + (2\pi\ell)^2]^{1/2}\,. \qquad (3.2.8)$$

For example, when $\ell = 1$, the critical value is $K_c = 2\pi$ so that for $K < K_c$, the accelerator mode is unstable and a domain exists in the phase space for $K > K_c$ when the accelerator mode is stable. An example of such an island is given in Fig. 3.2.1. Its phase portrait is presented on the torus: $(x, p) \in (0, 2\pi)$. The invariant closed curves inside the island correspond to the accelerated trajectories in the infinite

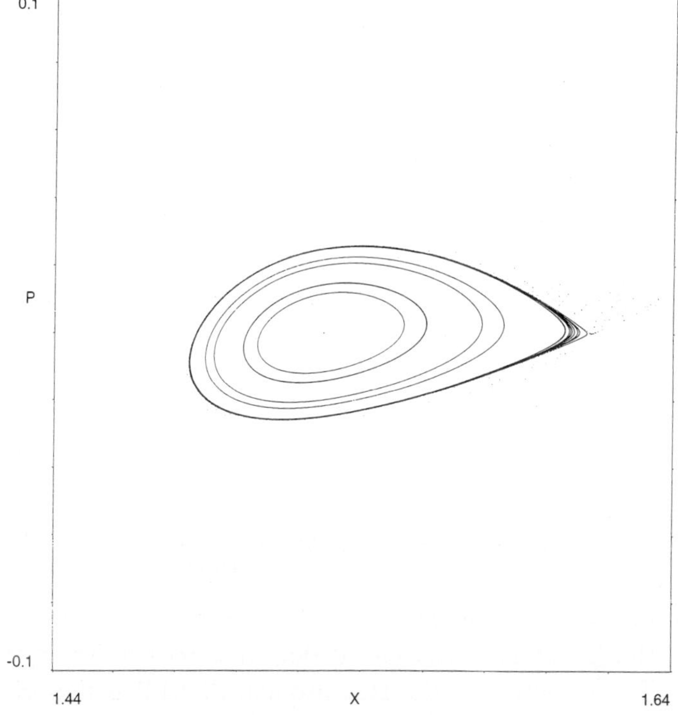


Fig. 3.2.1. Borning (collapsing) island for the standard map ($K = 6.2890$).

phase space. Any initial condition outside the island will correspond to the chaotic trajectory.

In writing the Hamiltonian description for the invariant curves inside the island, $K_c = 2\pi$ is retained for the sake of simplicity, $K = 2\pi + \Delta K$, $x = \pi/2 + \Delta x$ and $p = \Delta p$, where

$$\Delta K = K - K_c = K - 2\pi\,, \tag{3.2.9}$$

$$\Delta x = x - x^{(a)}\,, \quad \Delta p = p - p^{(a)}\,, \quad (\mathrm{mod}\,2\pi) \tag{3.2.10}$$

and $x^{(a)}, p^{(a)}$ belong to the accelerated trajectories (3.2.2) and (3.2.3) taken on the torus. This means that a trajectory $(x, p)$ in the vicinity of the trajectory with initial points $(x_0^{(a)}, p_0^{(a)}) = (\pi/2, 0)$ on the torus is taken into consideration. For terms up to the second order in $\Delta x$, $\Delta p$ and the first order in $\Delta K$, the equations can be derived from (3.2.1):

$$\begin{aligned} \Delta\dot{p} &= -\Delta K + \pi(\Delta x)^2 \\ \Delta\dot{x} &= \Delta p\,, \end{aligned} \tag{3.2.11}$$

where

$$\Delta\dot{p} \approx p_{n+1} - p_n\,, \quad \Delta\dot{x} \approx x_{n+1} - x_n \tag{3.2.12}$$

are replaced and the dot refers to the derivative with respect to dimensionless time. Equation (3.2.11) are of the Hamiltonian type with a canonical pair of variables $(\Delta p, \Delta x)$:

$$\begin{aligned} H^{(a)} &= \frac{1}{2}(\Delta p)^2 + \Delta K \cdot \Delta x - \frac{\pi}{3}(\Delta x)^3 \\ \Delta\dot{p} &= -\frac{\partial H^{(a)}}{\partial \Delta x}\,, \quad \Delta\dot{x} = \frac{\partial H^{(a)}}{\partial \Delta p}\,. \end{aligned} \tag{3.2.13}$$

It can be easily concluded from (3.2.13) that the island collapses at $\Delta K = 0$, that is, at $K = K_c$, and it does not exist if $\Delta K < 0$, that is, $K < K_c$. The structure of the Hamiltonian (3.2.13) is important since it defines a rescaling possibility in the form of

$$\begin{aligned} &\Delta x \to \xi(\Delta K)^{1/2}\,, \quad \Delta p \to \eta \cdot (\Delta K)^{3/4} \\ &H^{(a)}(\Delta x, \Delta p) \to H^{(a)}(\xi, \eta) \cdot (\Delta K)^{3/2}\,, \end{aligned} \tag{3.2.14}$$

where $(\xi, \eta)$ are rescaled variables. This case is further discussed in Section 3.4.[3]

A similar example is provided for the web-map $\hat{M}_4$ (1.3.12):

$$u_{n+1} = v_n\,, \quad v_{n+1} = -u_n - K \sin v_n\,. \tag{3.2.15}$$

For instance, the conditions

$$K_c = 2\pi\,, \quad u_0 = 3\pi/2\,, \quad v_0 = \pi/2 \tag{3.2.16}$$

---

[3]The existence of islands of the type indicated in Eqs. (3.2.13) is shown in [Me 96]. Connection between the islands and the accelerator mode is shown in [ZEN 97].

define the accelerator mode solution

$$u_{n+4}^{(a)} = u_n^{(a)} + 4\pi\,, \quad v_{n+4}^{(a)} = v_n^{(a)} - 4\pi\,. \tag{3.2.17}$$

As in (3.2.10), the deviations from the value $K_c = 2\pi$ and the initial trajectory (3.2.17) are introduced:

$$\Delta u_{n+4} = u_{n+4} - u_{n+4}^{(a)}\,, \quad \Delta v_{n+4} = v_{n+4} - v_{n+4}^{(a)}\,, \quad \Delta K = K - K_c\,. \tag{3.2.18}$$

The absence of the superscript (a) indicates an arbitrary trajectory (3.2.15) where the parameter $K$ and the initial conditions are close to $K_c$ and (3.2.16), respectively. After (3.2.15) has been iterated four times and expanded up to the second order terms in $\Delta u, \Delta v$ and the first order terms in $\Delta K$, it yields

$$\begin{aligned} \Delta u_{n+4} &= \Delta u_n - 2\Delta K + 2\pi(\Delta v_n)^2 \\ \Delta v_{n+4} &= \Delta v_n - 2\Delta K + 2\pi(\Delta u_{n+4})^2 \end{aligned} \tag{3.2.19}$$

or simply,

$$\begin{aligned} \Delta\dot{u} &= -\frac{1}{2}\Delta K + \frac{1}{2}\pi(\Delta v)^2 \\ \Delta\dot{v} &= -\frac{1}{2}\Delta K + \frac{1}{2}\pi(\Delta u)^2 \end{aligned}\,, \tag{3.2.20}$$

where the dot indicates a time derivative that appears after a replacement similar to (3.2.12) has taken place:

$$\begin{aligned} \frac{1}{4}(\Delta u_{n+4} - \Delta u_n) &\approx \Delta\dot{u}_n \\ \frac{1}{4}(\Delta v_{n+4} - \Delta v_n) &\approx \Delta\dot{v}_n \end{aligned}\,.$$

The system (3.2.20) can be written in the Hamiltonian form

$$\Delta\dot{u} = -\frac{\partial H^{(a)}}{\partial \Delta v}\,, \quad \Delta\dot{v} = \frac{\partial H^{(a)}}{\partial \Delta u} \tag{3.2.21}$$

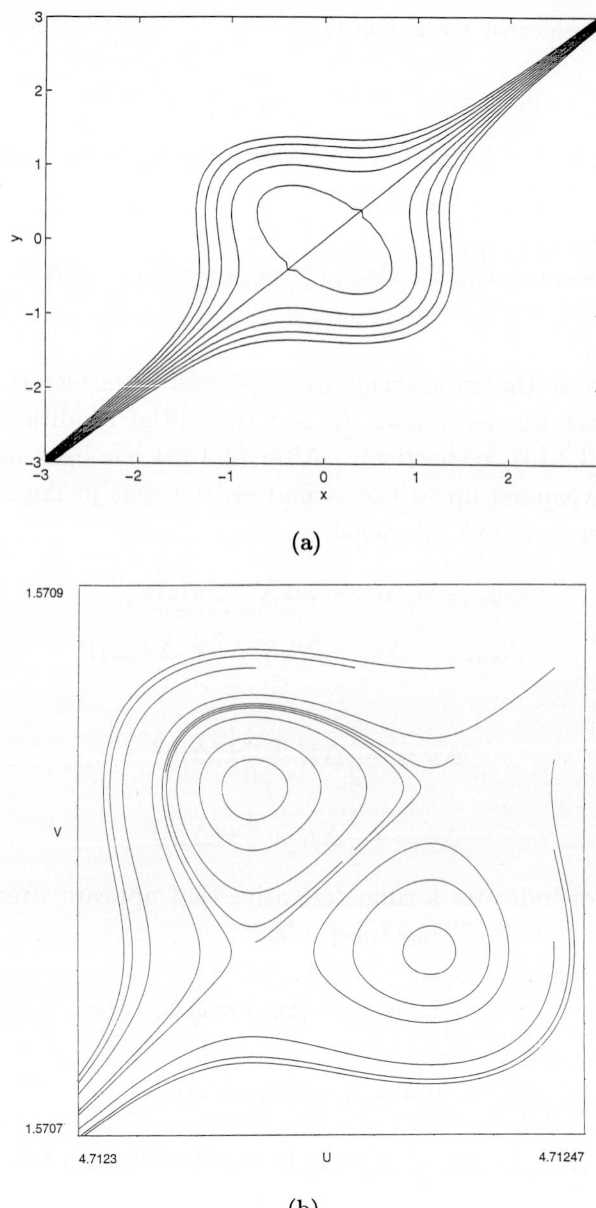


Fig. 3.2.2. Borning (collapsing) pair of islands for the web-map: (a) isolines of the model Hamiltonian; and (b) real phase portrait ($q = 4$, $K = 6.28318531$).

with

$$H^{(a)} = \frac{1}{2}\Delta K(\Delta v - \Delta u) - \frac{\pi}{6}[(\Delta v)^3 - (\Delta u)^3]. \tag{3.2.22}$$

The isolines for the Hamiltonian (3.2.22) and separatrices are presented in Fig. 3.2.2(a). The corresponding trajectories for map (3.2.15) and $K = 6.28318531$, which are very close to $K_c = 2\pi$, are given in Fig. 3.2.2(b). A coincidence is observed in the topology of the patterns. The double-island structure occurs only when

$$\Delta K = K - K_c > 0 \tag{3.2.23}$$

and collapses to zero, while at the same time preserving the topology imposed by the Hamiltonian $H^{(a)}$ in (3.2.22). The Hamiltonian of the collapsing islands satisfies a rescaling transform

$$\begin{aligned} &\Delta u \to \xi(\Delta K)^{1/2}, \quad \Delta v \to \eta(\Delta K)^{1/2} \\ &H^{(a)}(\Delta u, \Delta v) \to (\Delta K)^{3/2} H^{(a)}(\xi, \eta) \end{aligned} \tag{3.2.24}$$

which is similar to (3.2.14) for the accelerated mode of the standard map. Nevertheless, like the phase space topologies, the Hamiltonians (3.2.22) and (3.2.14) are different.[4]

## 3.3 Blinking Islands

Accelerator mode islands belong to a special category of island species. This is because the existence of accelerator modes is a specific property that occurs in only in some problems where the so-called ballistic type of motion takes place. The more typical type of islands are the "blinking" islands, which only collapse for the special value $K = K_c$ and have a different orientation for $K < K_c$ and $K > K_c$. An example of the collapse of such an island is presented in Fig. 3.3.1 for the standard map (3.2.1). It reveals two important features:

---

[4]The structure of the islands in (3.2.21) and (3.2.22) was obtained in [ZN 97] and [ZEN 97].

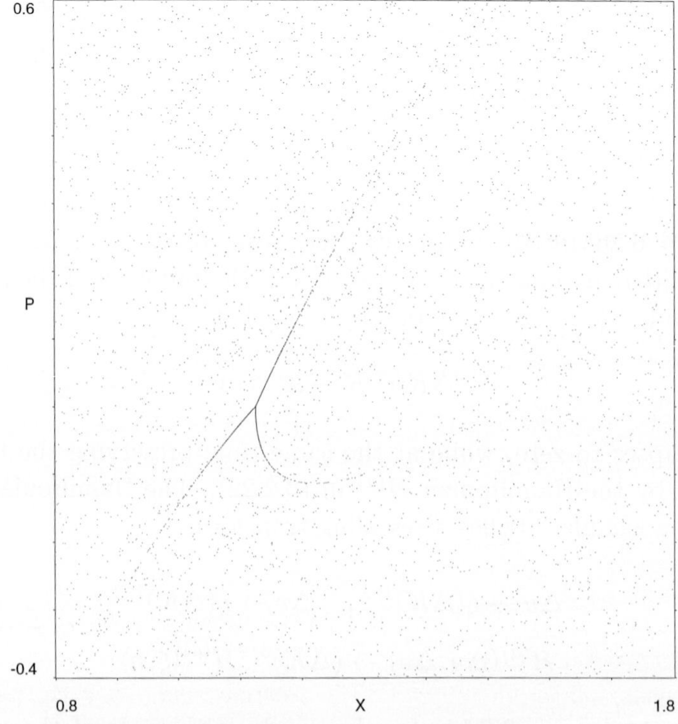


Fig. 3.3.1. Collapse of the blinking island for the standard map ($K = 6.962643$).

(i) The local symmetry of order three.

(ii) The presence of "scars", that is, regions of stagnated trajectories that can influence the transport properties of particles.

It is fairly easy to describe the bifurcation and collapse of an island using its symmetry property. The initial occurrence of an island in the stochastic sea is accompanied by the appearance of a stable elliptic point in the centre of the island with stable orbits around it. The symmetry of a characteristic Hamiltonian with orbits inside the island is of the third order, that is, the Hamiltonian should be of the third order polynomial in $\Delta x, \Delta p$. When it is stated that

$$H_0 = \frac{1}{2}(\Delta p)^2 + V(\Delta x, \Delta p; \Delta K), \tag{3.3.1}$$

the same type of variables near the value $K_c$ and the trajectory $(p_c, x_c)$ of the collapsed island are used:

$$\Delta p = p - p_c\,, \quad \Delta x = x - x_c\,, \quad \Delta K = K - K_c\,. \tag{3.3.2}$$

$(p_c, x_c)$ is the elliptic (central) point of the island and $V$ is the "potential" of the third order polynomial

$$V(\Delta x; \Delta p; \Delta K) = \prod_{j=1}^{3} (a_j \Delta x + b_j \Delta p + c_j \Delta K) \tag{3.3.3}$$

with some constants $a_j, b_j, c_j$ being of order one. The equations of motion are Hamiltonian

$$\Delta\dot{p} = -\frac{\partial H_0}{\partial \Delta x} \quad \Delta\dot{x} = \frac{\partial H_0}{\partial \Delta p}\,. \tag{3.3.4}$$

For small $\Delta K$, a triangle is formed by three separatrix lines, all of which intersect at a single point if $\Delta K = 0$ (Fig. 3.3.2). The island shrinks when $K$ passes through the value $K_c$ but reappears after the passing.[5]

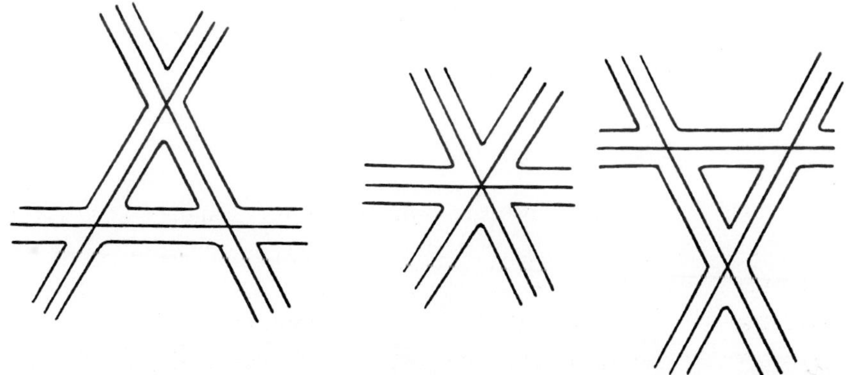

Fig. 3.3.2. Scheme of bifurcations with the collapse of the blinking island.

[5]The existence of the bifurcation presented in Fig. 3.3.2 can be established in a general way by considering the Birkhoff normal form for a Hamiltonian [A 89]. The numerical simulation found in Fig. 3.3.3 follows that derived in [ZEN 97].

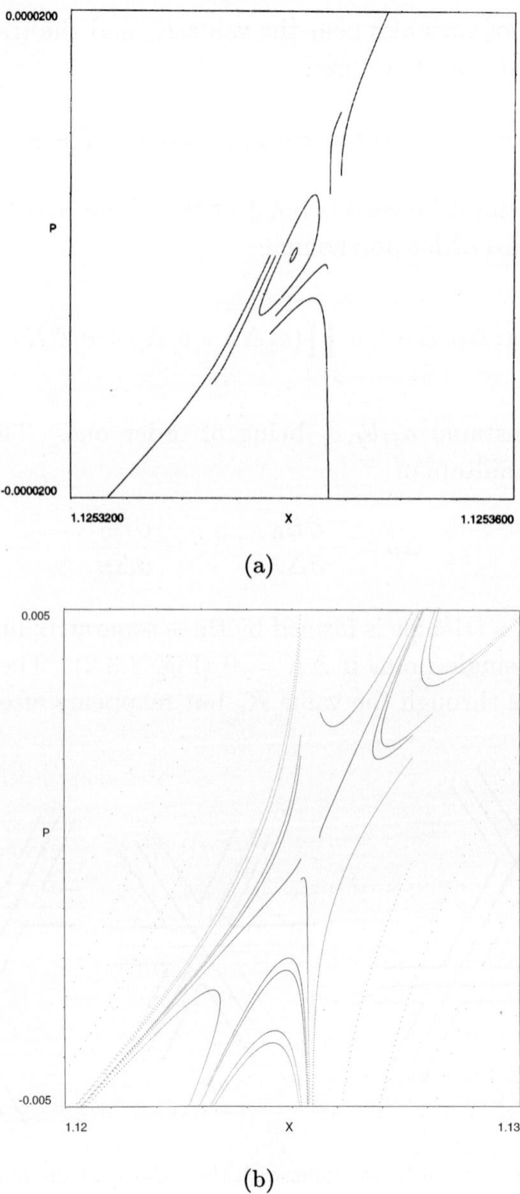


Fig. 3.3.3. The same as in Fig. 3.3.2 but for the standard map with: (a) $K = 6.962640$ (before the bifurcation); (b) $K = 6.962643$ (close to the bifurcation); and (c) $K = 6.962650$ (after the bifurcation).

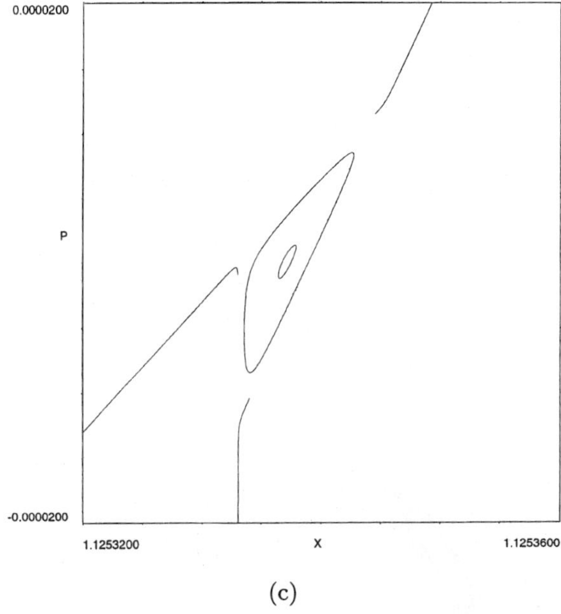


(c)

Fig. 3.3.3. (*Continued*)

The main feature of a blinking island is that although it is absent for a special value of the parameter $K = K_c$, it exists for $K \neq K_c$, as opposed to the accelerator islands which only exist for $K > K_c$. This phenomenon can be observed for the standard map. In the example given in Fig. 3.3.3, which shows a domain near the island that has been magnified strongly before and after the "blink", the parameter $K$ is taken near its critical value $K_c = 6.96264\ldots$. The replacement of $\Delta K \to -\Delta K$ in (3.3.3) leads to the rotation of the island by $\pi$. This is observed in Fig. 3.3.3 in the transition from "a" to "c". The existence of six (three stable and three unstable) separatrices is evident in Fig. 3.3.3(b). Figure 3.3.4 displays a basin near the island with trajectories that have been strongly trapped. Since they are very close to the stable/unstable manifolds, the trajectories slow down the motion. This is one possible explanation for the stickiness and anomalous transport of particles.

Similar blinking islands can also be found for the web-map and for the problem of the perturbed pendulum.

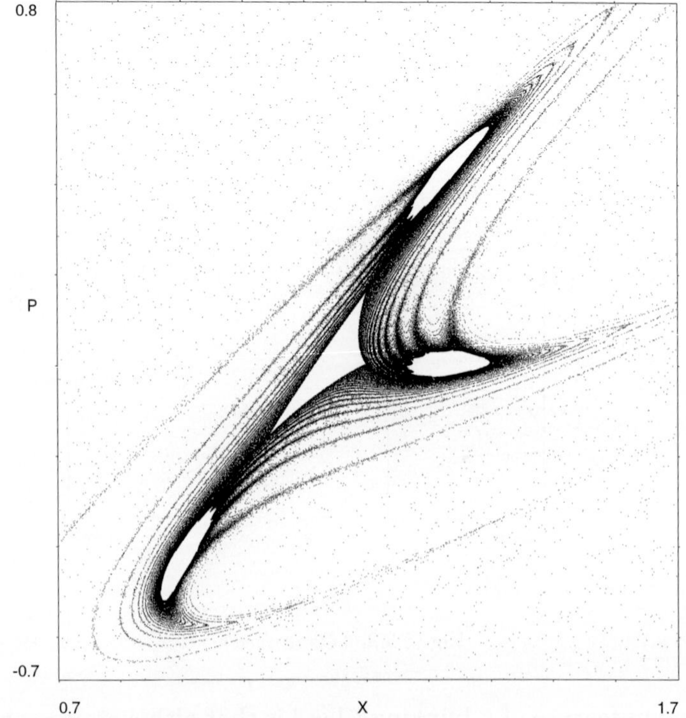


Fig. 3.3.4. Singular zone for the standard map ($K = 6.9110$).

## 3.4 Boundary Islands

Boundary islands live in the stochastic domain near the boundary of another island. Boundary islands surround the central, parental island by creating a boundary island chain (BIC) described in Section 2.4. Their appearance/disappearance depends on the shift of the boundary of the central island when a parameter changes. This means that the transformation of the boundary islands is described on the basis of boundary transformation. Nevertheless, it is extremely difficult to do so because of the fine structure of the boundary.

Imagine a central island embedded in (or surrounded by) the domain of chaotic motion with a boundary separating the area of chaos (see schematic picture in Fig. 3.4.1). One can then easily imagine a set of

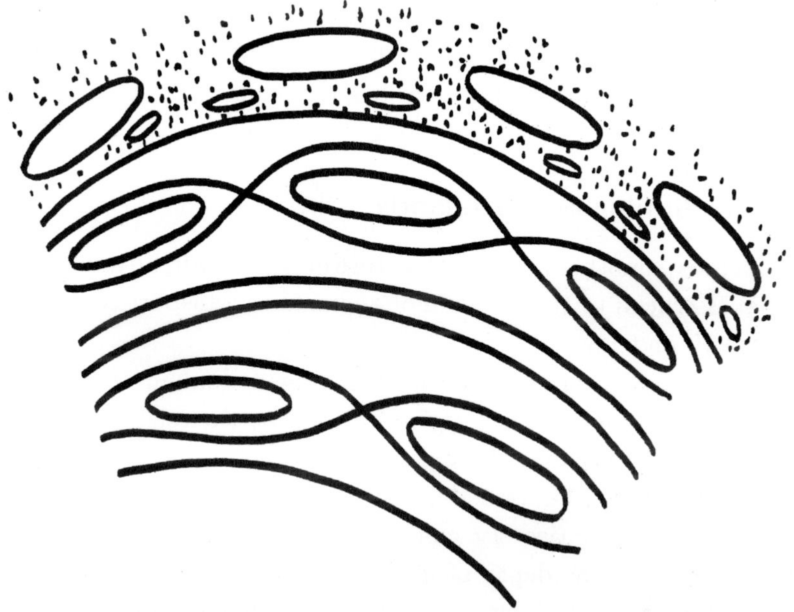

Fig. 3.4.1. Scheme of the island chains near a border of the stochastic sea.

chains of islands separated from the chaotic area by invariant curves inside the central island. Figure 3.4.1 shows two types of such chains known as two-chain and three-chain. One can determine the resonances related to these chains, and even for those related to the next chain (four-chain) located in the area of chaos. A description of this procedure is found in Sections 2.1 and 2.2. In fact, four-chain consists of only remnants of islands which can be fairly small. By changing a control parameter, say $K$, for the standard map so that the boundary of the chaotic area begins to move, a modification of the islands will occur. Numerous bifurcations influence what takes place near the boundary of chaos where many smaller islands can appear and disappear. There is no general description of the evolution of the pattern near the boundary of the area of chaos. Nevertheless, some special properties and configurations can be indicated, described and explored. One such case is considered in the next section.

A small but important comment is that the notion of boundary islands is a "geographic" one and can be applied to, amongst others, a thin stochastic layer, a stochastic web and a small island in the stochastic sea.[6]

## 3.5 Self-Similar Set of Islands

In this section, another possible structure of islands will be introduced — one that is crucial to the problem of anomalous transport considered in Chapters 10 and 11.

The probability of a trajectory entering a domain containing an island and its boundary layer is proportional to the relative area of the domain. This is because the distribution function of the trajectory's points outside the islands and their neighbourhood is uniform. The amount of time a trajectory can spend in the boundary layer of an island depends on the depth of its penetration into the layer, and this time can be extremely long. This trapping phenomenon is sometimes called "stickiness". Since its discovery, stickiness has remained a difficult puzzle in dynamical systems theory. It is linked to the existence of a very complex phase space topology near an island's boundary. The difficult part of this phenomenon is that the level of stickiness, or its characteristic trapping time, depends on an external parameter such as $K$ for the standard map or web-map. The stickiness can also be a lot stronger near some critical values of $K$. For example, in Fig. 3.3.4, the darkness means a very strong stickiness near the value of $K$ which corresponds to the bifurcation of three-islands separation. In Fig. 3.1.4(b), the stickiness occurs near the zero-measure phase space domains which correspond to unstable periodic trajectories of free, that is, non-scattered, motion.

The behaviour of the reasonance islands chain described in the previous section and shown in Fig. 3.4.1 has given rise to the possibility that the structure of the islands may possess the property of self-similarity. This property is described for the web-map (3.2.15).

[6] An example of a special configuration of islands is the "islands-around-islands" structure described in [M 86] and [M 92].

There are domains of chaotic motion in the phase space $(u, v)$, (mod $2\pi$) for all values of $K$. Beginning with the value $K_0 > 1$, there are two kinds of domains in non-chaotic motion: islands of periodic motion and islands of accelerator modes. An example is found in Fig. 3.1.2(a), where large-sized islands correspond to the periodic trajectories and small-sized islands correspond to the accelerator modes from (3.2.16) to (3.2.22). In other words, in the infinite phase space $(u, v)$, a trajectory is bounded if its initial condition is taken inside an island in the periodic motion, and it is unbounded when the initial condition is found inside an accelerator-mode island.

After allowing $\ell$ to be the condition for the occurrence of an accelerator-mode solution,

$$K > K_\ell = 2\pi\ell\,, \quad \ell = 1, 2, \ldots, \tag{3.5.1}$$

the accelerator-mode island comes into existence. It is described by the Hamiltonian (3.2.22) for the values $\Delta K = K - K_\ell$ which satisfy the condition

$$K \in (K'_\ell,\, K''_\ell)\,, \tag{3.5.2}$$

where $K'_\ell = K_\ell$ and $K''_\ell > K_\ell$ is defined by the value of the first bifurcation. In Fig. 3.5.1(a), we present the computer simulation for $\ell = 1$ and $K = 6.349972$. Due to the four-fold symmetry of map (3.2.15), there are four equivalent groups of accelerator-mode islands. A magnified view of the right-bottom group of islands from Fig. 3.5.1(a) is presented in Fig. 3.5.1(b), where different initial conditions are taken both inside and outside the islands. One can see a resonance seven-island chain inside the top island and an eight-island chain inside the stochastic sea near the boundary of chaos. We refer to this chain as a boundary island chain. Small variations in $K$ will change the number of islands in the boundary chain. In Fig. 3.5.1(b), this number is eight. However, one can easily move the boundary of chaos inside so that the inner chain of seven islands becomes the outer boundary chain. It is possible to either increase or decrease the number of islands in the boundary chain.

When $K_{\ell;m_1}$ is a value of $K$ which corresponds to the $m_1$ chain in the boundary layer of the $\ell$-order accelerator mode, the number $m_1$ exists if

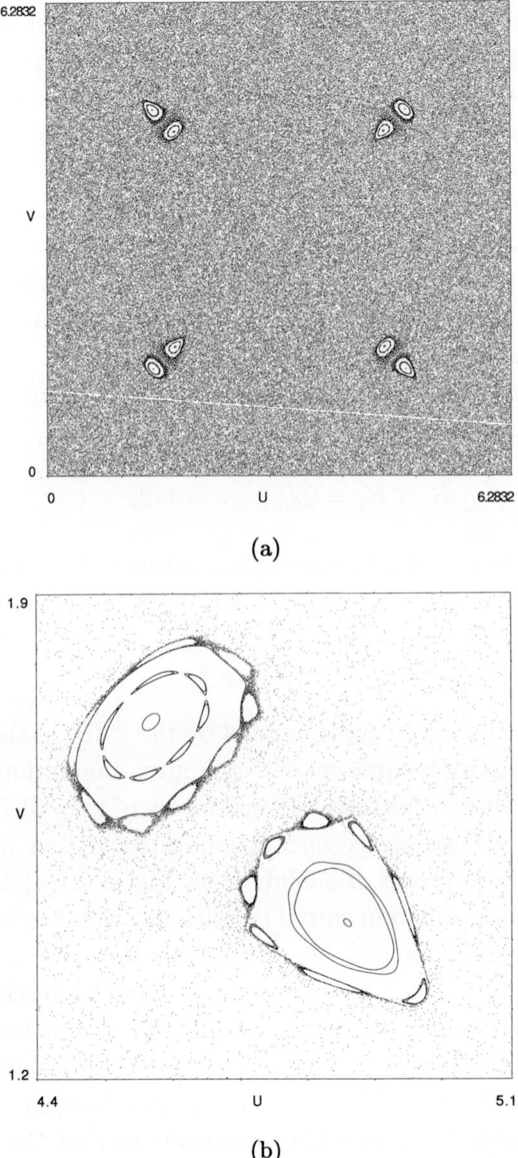


Fig. 3.5.1. Self-similar structure of islands for the web map with $K = 6.349972$: (a) the phase space with islands of the accelerator mode; (b) magnification of the bottom right part of (a); (c) magnification of the top right island of (b); and (d) magnification of the bottom left island of (c).

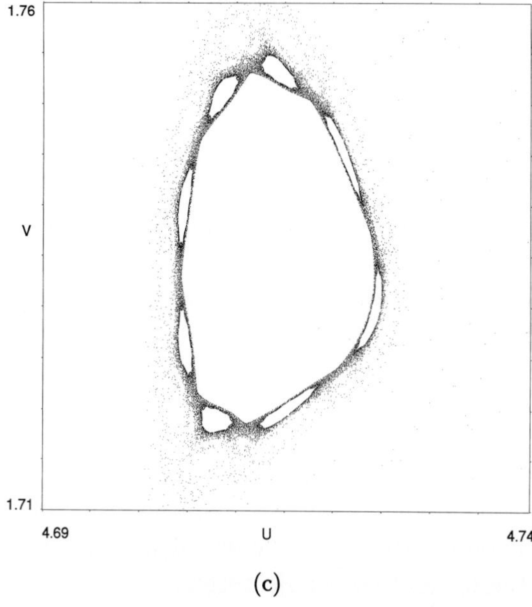


(c)

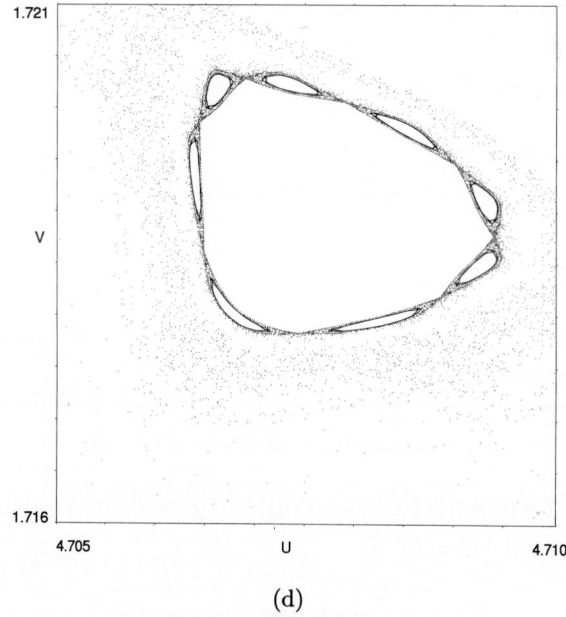


(d)

Fig. 3.5.1. (*Continued*)

$$K \in (K'_{\ell;m_1}, K''_{\ell;m_1}) \subset (K'_\ell, K''_\ell)\,. \tag{3.5.3}$$

The number $m_1$ refers to the resonance order of the first generation chain in the boundary layer of the parent island. Long time computation displays the stickiness for first generation islands and one of them (top right of Fig. 3.5.1(b)) is shown in Fig. 3.5.1(c). When magnified, the first generation island reveals the islands in the second generation (Fig. 3.5.1(d)). When $m_2$ is the number of islands in the second generation chain, a number $m_2$ exists which is analogous to (3.5.3) if

$$K \in (K'_{\ell;m_1m_2}, K''_{\ell;m_1m_2}) \subset (K'_{\ell;m_1}, K''_{\ell;m_2})\,. \tag{3.5.4}$$

The second generation boundary chain is also sticky. Thus, one can continue the process for subsequent generations. This process can reach infinity, that is, the value $K_{\ell;m_1m_2}\cdots$ specifies an infinite number of sequences in the boundary chain islands $(m_1m_2\ldots)$. This result can be formulated in the form of two conjectures.

**Conjecture 1.** Such values for $K$ exist which generate a chain

$$K \in (K'_{\ell;m_1\ldots m_n}, K''_{\ell;m_1\ldots m_n}) \subset \cdots \subset (K'_{\ell;m_1}, K''_{\ell;m_1})\,, \quad (n \to \infty) \tag{3.5.5}$$

for a given "word"

$$\{m_j\}_1^n = \{m_1, \ldots, m_n\} \tag{3.5.6}$$

with some reasonable restrictions on the value of $m$. For example, $m_j \geq 3$ and $m_1 \sim m_2 \sim \cdots \sim m_n$.

**Conjecture 2.** Such a value for $K_{\ell;m_s}$ exists in which all the chains have the same $m$ starting from some $j$-th:

$$m_j = m_{j+1} = \cdots \equiv m_s\,, \quad (1 \leq j)\,. \tag{3.5.7}$$

The case for (3.5.6) with $j = 1$ and $m_s = 8$ is presented in Figs. 3.5.1(a)–(d) for the special value of $K$:

$$K_{\ell=2;m_s=8} = 6.349972\cdots = K_{an}^w\,, \tag{3.5.8}$$

where $K_{an}^{w}$ refers to the anomalous value of $K$ for the web-map. The larger the word in (3.5.5), the smaller the domain in $(K', K'')$ becomes for the corresponding value of $K$. To derive more generations with the proliferation number eight, one can determine more digits in (3.5.8). We can consider an alternating sequence for values $m_1, m_2$: $\{m_1 m_2 m_1 m_2 \ldots\}$ or for more complicated sequences, and the values of $K$ that correspond to these sequences.

The boundary layer of a parental island is termed the singular zone of the phase plane. The hierarchy of island chains described above is a self-similar structure if condition (3.5.7) is fulfilled. A situation can also exist, called quasi-self-similarity, when all or almost all $m_j$ from the word $\{m_j\}_1^\infty$ are close to some value $m$.[7]

The self-similarity of the island chain can be presented in a more explicit way if we describe the different generations by the number of islands $q_k$ in the $k$-th generation chain, the area $\Delta S_k$ of an island in the same chain, and the period $T_k$ of the last invariant curve inside an island in $k$-generation chain. The data are shown in Table 3.5.1 where an auxiliary variable,

$$\delta S_k = q_k \Delta S_k \, , \tag{3.5.9}$$

is used to present the area of a full $k$-th generation chain. This is because in terms of the area-preserving dynamics, all the islands have the same area if they belong to the same chain.

The first column in Table 3.5.1 refers to the order $k$ for the corresponding generation and all the data are derived for the value $K = K_{an}^{w}$ in (3.5.8). The total number of islands of the type considered in the singular zone can be expressed as

$$N_n = q_1 \ldots q_n = \lambda_q^n \, , \tag{3.5.10}$$

[7]This situation was first described in the article [ZSW 93] where the problem of passive particle advection was considered for the so-called ABC model of helical flow. Exact self-similarity was first found for the web-map, which is more simple and symmetric [ZN 97], followed by the standard map [ZEN 97, BKWZ 97]. The last reference includes a variety of cases on the standard map.

Table 3.5.1.

| $k$ | $q_k$ | $T_k$ | $T_k/T_{k-1}$ | $\Delta S_k$ | $\Delta S_k/\Delta S_{k-1}$ | $\delta S_k$ | $\delta S_k/\delta S_{k-1}$ |
|---|---|---|---|---|---|---|---|
| 0 | 1 | 16.4 | – | 0.436 | – | 0.436 | – |
| 1 | 8 | 131.8 | 8.04 | $5.24^{-3}$ | $1.20 \times 10^{-2}$ | $4.19 \times 10^{-2}$ | 0.0961 |
| 2 | 8 | 1049 | 7.96 | $5.30^{-5}$ | $1.01 \times 10^{-2}$ | $3.39 \times 10^{-3}$ | 0.0809 |
| 3 | 8 | 8420 | 8.02 | $5.32^{-7}$ | $1.00 \times 10^{-2}$ | $2.72 \times 10^{-4}$ | 0.0802 |

where the scaling parameter $\lambda_q$ is introduced. In the case of (3.5.8), it is $\lambda_q = 8$. Two other scaling parameters, $\lambda_T$ and $\lambda_S$, exist such that

$$\begin{aligned} T_{k+1} &= \lambda_T T_k\,, \qquad \lambda_T > 1 \\ \delta S_{k+1} &= \lambda_S \delta S_k\,, \quad \lambda_S < 1 \end{aligned} \tag{3.5.11}$$

are responsible for the time-space self-similarity in the singular zone. From Table 3.5.1,

$$\lambda_T \approx 8\,, \quad \lambda_S \approx 0.08\,. \tag{3.5.12}$$

The self-similarity of the Lyapunov exponents is therefore

$$\sigma_{n+1} = \lambda_\sigma \sigma_n\,, \tag{3.5.13}$$

since no other independent process with a different time scale exists. It is important that the singular zone has the property of stickiness, that is, a trajectory spends a fairly long time in the zone. As a result of such behaviour, trajectories are entangled via the hierarchy of islands and the self-similarity properties (3.5.10) to (3.5.13) impose a specific large-scale behaviour on the trajectories. Chapter 11 considers the consequences of the existence of such properties in the singular zone.

Regarding the number of independent scaling parameters $\lambda_j$, it is evident that for a simplified situation,

$$\lambda_q = \lambda_T \tag{3.5.14}$$

and

$$\lambda_T \lambda_\sigma = 1\,. \tag{3.5.15}$$

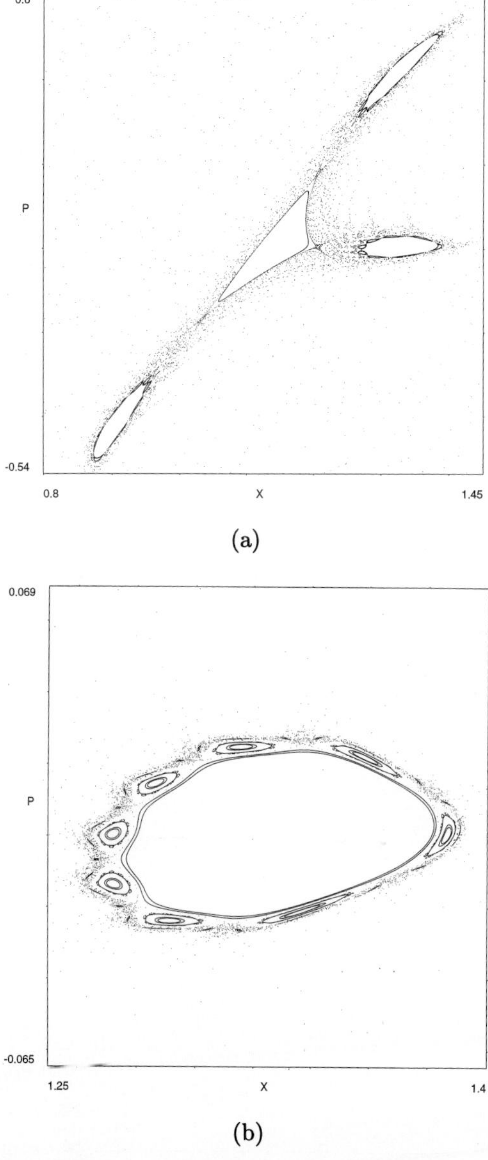


Fig. 3.5.2. Self-similar structure of islands for the standard map with $K = 6.908745$: (a) the phase space with islands of the accelerator mode; (b) magnification of the right island of (a); (c) magnification of the top left island of (b); and (d) magnification of the bottom island of (c).

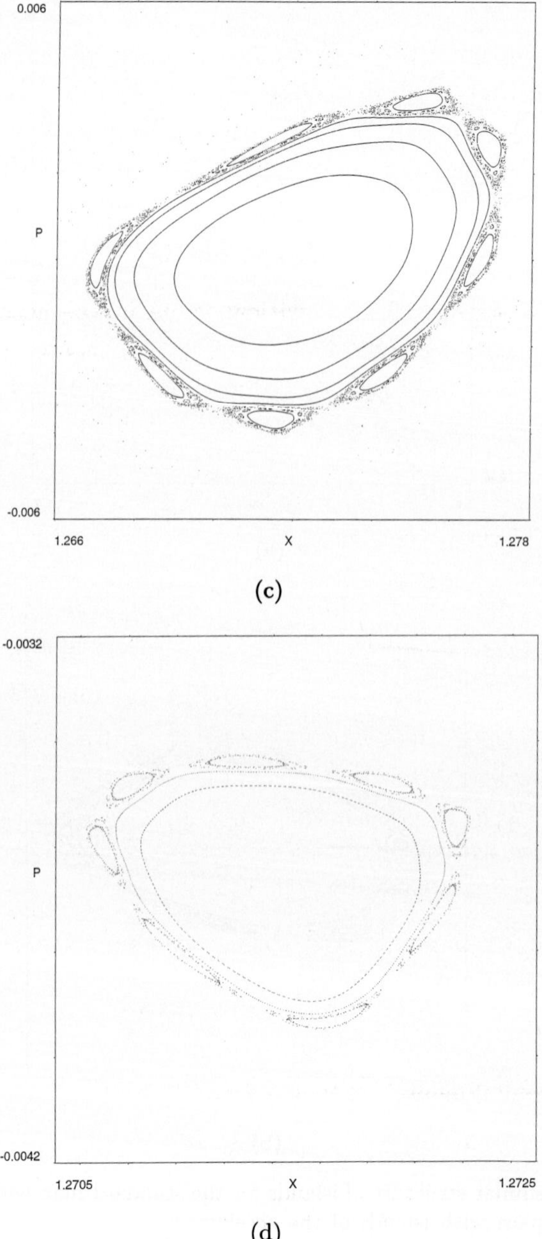


(c)

(d)

Fig. 3.5.2. (*Continued*)

Nevertheless, there could be a violation of the relations (3.5.14) and (3.5.15) when more complicated words $\{m_j\}$ are considered.

Another example of the self-similarity of islands in the singular zone for the standard map (3.2.1) is given below. Again, it corresponds to the accelerator mode solution (3.2.2), (3.2.3) and (3.2.9) to (3.2.13).[8] The value

$$K = K_{an}^S = 6.908745 \tag{3.5.16}$$

is close to $K = 2\pi$, which is a critical value in the accelerator mode solution. The set of islands' chains for four subsequent generations is shown in Figs. 3.5.2(a)–(d). In fact, more generations are derived from the simulation and they are all presented in Table 3.5.2.

Table 3.5.2.

| $k$ | $q$ | $T_k$ | $T_k/T_{k-1}$ | $\Delta S_k$ | $\Delta S_k/\Delta S_{k-1}$ | $\delta S_k$ | $\delta S_k/\delta S_{k-1}$ |
|---|---|---|---|---|---|---|---|
| 0 | 1 | 2.97 | – | $1.16 \times 10^{-2}$ | – | $1.16 \times 10^{-2}$ | – |
| 1 | 3 | 23.4 | 7.89 | $3.77 \times 10^{-3}$ | 0.324 | $1.13 \times 10^{-2}$ | 0.971 |
| 2 | 8 | 187 | 7.97 | $4.51 \times 10^{-5}$ | $1.20 \times 10^{-2}$ | $1.08 \times 10^{-3}$ | 0.0959 |
| 3 | 8 | 1500 | 8.05 | $4.86 \times 10^{-7}$ | $1.08 \times 10^{-2}$ | $0.934 \times 10^{-4}$ | 0.0862 |
| 4 | 8 | 12000 | 7.97 | $4.93 \times 10^{-9}$ | $1.01 \times 10^{-2}$ | $0.758 \times 10^{-5}$ | 0.0812 |
| 5 | 8 | 95800 | 8.00 | $4.81 \times 10^{-11}$ | $0.98 \times 10^{-2}$ | $0.592 \times 10^{-6}$ | 0.0781 |

From Table 3.5.2, one derives, with satisfactory accuracy,

$$\lambda_T = 7.98\,, \quad \lambda_S = 0.085\,, \tag{3.5.17}$$

and the sequence of islands is defined by the word

$$\{m_j\} = \{3, 8, 8, 8, 8 \ldots\}\,.^9 \tag{3.5.18}$$

[8]See footnote [7].

[9]The data in Tables 3.5.1 and 3.5.2 were extracted from [ZN 97] and [ZEN 97].

## 3.6 General Comments About the Islands

The reason why so much attention has been paid to an understanding of the islands' structure in this chapter is because the different islands correspond to different physical processes responsible for their origin. An island in the stochastic sea refers to the domain of initial conditions that generates stable trajectories. Therefore, information about the islands allows one to control the chaos. There is another reason, though a less apparent one: islands near their bifurcation states and self-similar sets of sticky islands strongly influence the type of kinetics present in the phase space. We can compare the structure of the phase space with the configuration space of matter near its critical point. Anomalous transport phenomena cannot be understood without a detailed description of the singular zone. This statement will become clearer in later chapters.

## Conclusions

1. The part of the phase space that has chaotic dynamics is called the stochastic sea. Inside the "sea" are islands of stability. A chaotic trajectory cannot penetrate the islands and a regular trajectory from an island is not able to escape from it.
2. A set of islands has a complex topology which can be modified by a control parameter. In particular, for some values of the control parameter, an island can either collapse or be born. Generally, in typical systems, chaos does not exist without any islands. The structure of islands which are on the verge of collapse can be studied with the help of an effective Hamiltonian. Two possible models of such a collapse include:
   (a) For an accelerator mode island in the standard map, the effective Hamiltonian is:
   $$H_{\text{ef}}(\Delta p, \Delta x) = \frac{1}{2}(\Delta p)^2 + \Delta K \Delta x - \frac{\pi}{3}(\Delta x)^3 .$$
   (b) For an accelerator mode island in the web-map, the effective Hamiltonian is:

$$H_{\rm ef}(\Delta u, \Delta v) = \frac{1}{2}\Delta K(\Delta v - \Delta u) - \frac{\pi}{6}[(\Delta v)^3 - (\Delta u)^3]\,.$$

3. A strip near the boundary of an island is called the boundary layer or singular zone. Many islands are found in the boundary layer. After selecting a set of islands that form a chain around the parent island, one can then select a new (second) chain of islands surrounding the island in the first chain and so forth. The sequence of islands from different generations can form a self-similar hierarchical structure when the special values for the control parameter are chosen. This structure is very important in understanding chaotic transport in the stochastic sea.

4. The hierarchical structure of the islands in the singular zone can explain the origin of stickiness in trajectories located at the islands' boundaries.

*Chapter 4*

# NONLINEARITY VERSUS PERTURBATION

## 4.1 Beyond the KAM-Theory

This chapter continues with a consideration of 1 1/2 degrees of freedom systems, that is, a time-dependent perturbation in a system with one degree of freedom. The results can be extended to systems with two degrees of freedom. The KAM-theory guarantees the persistence of invariant tori (curves) within the limits of the small perturbation

$$\epsilon \to 0 \tag{4.1.1}$$

with the non-degeneracy condition being

$$\left| \frac{\partial^2 H_0(I)}{\partial I_k \partial I_\ell} \right| \neq 0 \, . \tag{4.1.2}$$

The system to be considered is a general one described by the Hamiltonian

$$H = H_0(I) + \epsilon V(I, \vartheta, t) \, , \tag{4.1.3}$$

where $I = (I_1, \ldots, I_N)$ is the action-vector, $\vartheta = (\vartheta_1, \ldots, \vartheta_N)$ is the phase vector, and $\epsilon$ is the dimensionless parameter of the perturbation. The form of the Hamiltonian (4.1.3) shows that its unperturbed part, $H_0$, describes an integrable system since all the actions of $I_j$ are integrals of motion in the absence of perturbation. This is correct for the unperturbed sytstem, where $N = 1$ because all time-independent systems with one degree of freedom are integrable. The non-degeneracy condition (4.1.2) takes the form for $N = 1$ as

$$\frac{d^2 H_0(I)}{dI^2} = \frac{d\omega(I)}{dI} \neq 0\,, \tag{4.1.4}$$

where the nonlinear frequency

$$\omega(I) = \frac{dH_0(I)}{dI} \tag{4.1.5}$$

is introduced for the unperturbed motion. Condition (4.1.4) therefore concludes that the unperturbed system should be nonlinear for the results of the KAM-theory to be applicable.

From the physical point of view, condition (4.1.1) means that $\epsilon$ should be fairly small, or $\epsilon < \epsilon_0$. However, condition (4.1.2) has no meaning since it does not indicate a critical value, $\alpha_0$, for the dimensionless nonlinear parameter

$$\alpha = \left|\frac{d\omega}{dI}\right| \frac{I}{\omega}\,, \quad (\omega \neq 0)\,, \tag{4.1.6}$$

such that fairly small $\alpha < \alpha_0$ should be considered as a violation of condition (4.1.2). It would be more accurate to write the condition of validity in the KAM-theory as

$$\epsilon < C\alpha^{\delta}\,, \quad (\epsilon, \alpha \to 0) \tag{4.1.7}$$

with some $\delta > 0$ and constant $C$ independent (or slightly dependent on) of $\epsilon$. A situation which differs from that in the KAM-theory occurs when the inequality (4.1.7) is not valid. This is the subject of discussion in this chapter.

This problem can be described more specifically by using the models of perturbed pendulum and perturbed oscillator introduced in Sections 1.4 and 1.5. In considering the Hamiltonians

$$H = \frac{1}{2}\dot{x}^2 - \omega_0^2 \cos x + \epsilon \frac{\omega_0^2}{k^2} \cos(kx - \nu t) \tag{4.1.8}$$

for the perturbed pendulum and

$$H = \frac{1}{2}\dot{x}^2 + \frac{1}{2}\omega_0^2 x^2 + \epsilon\frac{\omega_0^2}{k^2}\cos(kx - \nu t) \tag{4.1.9}$$

for the perturbed oscillator, their equations of motion are, respectively,

$$\ddot{x} + \omega_0^2 \sin x = \epsilon\frac{\omega_0^2}{k}\sin(kx - \nu t) \tag{4.1.10}$$

and

$$\ddot{x} + \omega_0^2 x = \epsilon\frac{\omega_0^2}{k}\sin(kx - \nu t)\,. \tag{4.1.11}$$

The effect of the perturbation for a linear oscillator is at its strongest in the resonant case

$$\nu = n_0\omega_0\,, \tag{4.1.12}$$

which has an integer $n_0$. Unless another condition is mentioned, this assumption is used below. The amplitude of a linear oscillator grows linearly with time until it is interrupted by nonlinearity. The latter is induced by the same perturbation.

It is evident that system (4.1.8), but not (4.1.9), satisfies the condition of non-degeneracy in (4.1.4). At the same time, the expansion

$$\cos x = 1 - x^2/2 + x^4/24 \cdots \tag{4.1.13}$$

endows (4.1.8) with the same Hamiltonian as (4.1.9) for a fairly small nonlinear term $x^4$. This gives rise to the following questions: how small should $\epsilon$ be so that the KAM-theory can be applied in (4.1.8)? And how should one consider (4.1.9) which is a non-KAM case? The answer is found in the following sections.

## 4.2 Web-Tori

When considering the perturbed oscillators (4.1.9) and (4.1.11), the polar co-ordinates are

$$x = \rho\sin\phi\,; \quad \dot{x} = \omega_0\rho\cos\phi\,. \tag{4.2.1}$$

The following expansion,

$$\cos(kx - \nu t) = \cos(k\rho \sin\phi - \nu t) = \sum_m J_m(k\rho)\cos(m\phi - \nu t)\,, \quad (4.2.2)$$

is also used, where $J_m$ is the Bessel function. With the new variables described in (4.2.1), the Hamiltonian (4.1.9) becomes

$$H = \frac{1}{2}\omega_0^2\rho^2 + \frac{1}{k^2}\epsilon\omega_0^2 \sum_m J_m(k\rho)\cos(m\phi - \nu t)\,. \quad (4.2.3)$$

A term with $m = n_0$ is then singled out:

$$\begin{aligned} H = {} & \frac{1}{2}\omega_0^2\rho^2 + \frac{1}{k^2}\epsilon\omega_0^2 J_{n_0}(k\rho)\cos(m\phi - \nu t) \\ & + \frac{1}{k^2}\epsilon\omega_0^2 \sum_{m\neq n_0} J_m(k\rho)\cos(m\phi - \nu t)\,. \end{aligned} \quad (4.2.4)$$

This is followed by the introduction of new action-angle variables:

$$I = \omega_0\rho^2/2n_0\,, \quad \theta = n_0\phi - \nu t\,. \quad (4.2.5)$$

A new Hamiltonian is then written as

$$\tilde{H} = H - \nu I\,, \quad (4.2.6)$$

where $H$ is expressed as a function of $(I, \theta)$. The use of a direct calculation ensures that the equations

$$\dot{I} = -\frac{\partial \tilde{H}}{\partial \theta}\,, \quad \dot{\theta} = \frac{\partial \tilde{H}}{\partial I} \quad (4.2.7)$$

are equivalent to the equation of motion (4.1.11). By substituting (4.2.5) in (4.2.4) and (4.2.6), this yields

$$\begin{aligned} \tilde{H} = {} & (n_0\omega_0 - \nu)I + \frac{1}{k^2}\epsilon\omega_0^2 J_{n_0}(k\rho)\cos\theta \\ & + \frac{1}{k^2}\epsilon\omega_0^2 \sum_{m\neq n_0} J_m(k\rho)\cos\left[\frac{m}{n_0}\theta - \left(1 - \frac{m}{n_0}\right)\nu t\right], \end{aligned} \quad (4.2.8)$$

where $\rho$ is introduced to obtain a more compact notation. According to (4.2.5), $\rho$ is

$$\rho = (2n_0 I/\omega_0)^{1/2} . \tag{4.2.9}$$

Thus, the expression $\tilde{H} = \tilde{H}(I, \theta; t)$ is the Hamiltonian with respect to the new canonical variables $(I, \theta)$. It can also be written as

$$\tilde{H} = \tilde{H}_0(I, \theta) + \tilde{V}(I, \theta; t) , \tag{4.2.10}$$

where in accordance with (4.2.8),

$$\begin{aligned} \tilde{H}_0(I, \theta) &= (n_0\omega_0 - \nu)I + \frac{1}{k^2}\epsilon\omega_0^2 J_{n_0}(k\rho)\cos\theta , \\ \tilde{V}(I, \theta; t) &= \frac{1}{k^2}\epsilon\omega_0^2 \sum_{m \neq n_0} J_m(k\rho)\cos\left[\frac{m}{n_0}\theta - \left(1 - \frac{m}{n_0}\right)\nu t\right] \end{aligned} \tag{4.2.11}$$

and expression (4.2.9) is used for $\rho$.

Turning now to the resonance case (4.1.12), the expression (4.2.11) for the Hamiltonian part $\tilde{H}_0$ takes the following form:

$$\tilde{H}_0 = \frac{1}{k^2}\epsilon\omega_0^2 J_{n_0}(k\rho)\cos\theta = \frac{1}{k^2}\epsilon\omega_0^2 J_{n_0}\left[k\left(\frac{2n_0 I}{\omega_0}\right)^{1/2}\right]\cos\theta . \tag{4.2.12}$$

In performing a preliminary analysis of the dynamic system that emerges in a resonance case, it is noted that the two terms in the Hamiltonian (4.2.10), $\tilde{H}_0$ and $\tilde{V}$, are proportional to $\epsilon$. Hence, the stationary part of the Hamiltonian, which is time-independent, is affected by the time-dependent part which is considered a perturbation. The fully transformed Hamiltonian $\tilde{H}$ vanishes at $\epsilon \to 0$. This is a new feature of the problem which has not yet been discussed.

It is possible to let $\tilde{H}_0(I, \theta)$ be an unperturbed part of the Hamiltonian $\tilde{H}$ and $\tilde{V}(I, \theta, t)$ for a perturbation. The phase portrait for the Hamiltonian (4.2.12), presented in Fig. 4.2.1, corresponds to the $(x, \dot{x})$ plane if a time instant is fixed. Separatrices form a net on the plane $(x, \dot{x})$ in the form of a spider-web with the number of rays being $2n_0$ and

Fig. 4.2.1. Phase portrait of system with the separatrix net in the form of a web ($n_0 = 4$).

the rotational symmetry at an angle of $\alpha = \pi/n_0$. The singular points of the system can be found from the following equations:

$$\frac{\partial \tilde{H}_0}{\partial I} = 0\,, \quad \frac{\partial \tilde{H}_0}{\partial \theta} = 0\,. \tag{4.2.13}$$

If $\tilde{H}_0$ is substituted by expression (4.2.12), a set of hyperbolic points $(\rho_h, \theta_h)$

$$J_n(k\rho_h) = 0\,, \quad \theta_h = \pm\pi/2 \tag{4.2.14}$$

and elliptic points $(\rho_e, \theta_e)$

$$J'_{n_0}(k\rho_e) = 0\,, \quad \theta_e = 0, \pi \tag{4.2.15}$$

is derived.

A family of separatrices is formed by $2n_0$ rays and circumferences with the radii $\rho_h^{(s)}$ crossing the rays at points which are solutions to the equations in (4.2.14). Inside the cells of the web, motion occurs along the closed orbits around the elliptic points located at the centre of the cells.

There is a principal difference between the motion described by the Hamiltonian (4.2.12) and the motion of a nonlinear pendulum. In the former, the web of separatrices covers the entire space with a regular set of finite meshes and the particle motion along the web can occur in a radial direction. In the latter, however, any motion in the radial direction is ruled out. This is a typical property in nonlinear cases where the non-degeneracy condition (4.1.2) or (4.1.4) holds.

However, radial motion, which is non-zero on average, is only possible along separatrices. In the vicinity of hyperbolic points, the radial motion is frozen. Hence there is no progress along the radius for the Hamiltonian $\tilde{H}_0$ (4.2.12) unless the perturbation $\tilde{V}$ in (4.2.11) is taken into account. As a result, the separatrices in the unperturbed form would disappear and be replaced by channels with finite width and stochastic dynamics in them. Before discussing this phenomenon in greater detail, a consideration of the structure of the unperturbed web-tori is given.[1] For the sake of simplicity, the regions which are sufficiently far from the centre of the web are considered, that is, with

$$k\rho \gg 1 \tag{4.2.16}$$

and the use of the asymptotics of the Bessel function:

$$J_n(k\rho) \sim \left(\frac{2}{\pi k\rho}\right)^{1/2} \cos(k\rho - \frac{1}{2}\pi n - \frac{1}{4}\pi)\,. \tag{4.2.17}$$

A cell in the web is singled out for a description of the family of unperturbed trajectories found inside it (Fig. 4.2.2). When $\rho_0$ is an elliptic

[1]The notion of tori (or web-tori) was used even though the Poincaré map consists of invariant curves and not tori. This is because the motion along the $z$ axis, that is, along the applied magnetic field, is not taken into consideration. When we do allow for the motion along $z$, the closed curves inside the cells of the web represent sections of the corresponding invariant tori. Also, a time axis can be used instead of the $z$ axis.

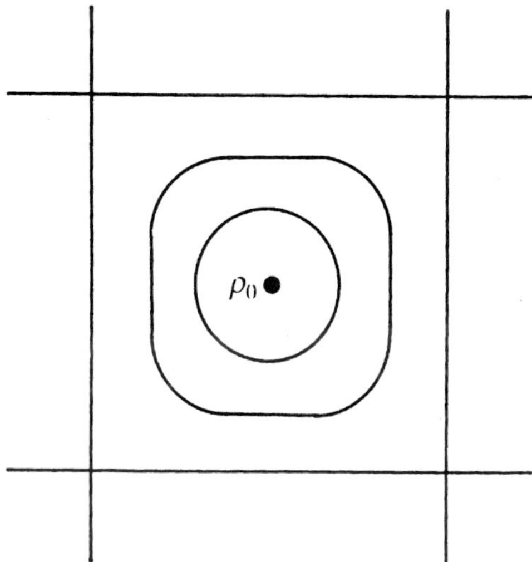


Fig. 4.2.2. Cell of the web.

point in the centre of the cell, then, according to (4.2.15) and (4.2.17), there are two possibilities:

$$k\rho_0 - \frac{1}{2}\pi n_0 - \frac{1}{4}\pi = 0\,, \quad k\rho_0 - \frac{1}{2}\pi n_0 - \frac{1}{4}\pi = \pi\,. \tag{4.2.18}$$

Using the expansion (4.2.17) and the condition (4.2.18), (4.2.12) is rewritten as

$$\tilde{H}_0 = -\sigma\epsilon\frac{\omega_0^2}{k^2}\left(\frac{2}{\pi k\rho_0}\right)^{1/2}\cos k\tilde{\rho}\cos\theta\,, \quad \sigma = \pm 1\,, \tag{4.2.19}$$

where $\tilde{\rho} = \rho - \rho_0$. The different signs also correspond to the different directions of rotations around the elliptic point as well as the different co-ordinates of the elliptic points. This depends on the value found on the right-hand side of (4.2.18).

An analysis of the trajectories determined by the Hamiltonian (4.2.19) shows that the radial size of the separatrix cell is of the order $2\pi/k$. Therefore, $\max\Delta\rho = 2\pi/k$ and according to the inequality (4.2.16),

$$|\tilde{\rho}| = |\rho - \rho_0| \ll \rho_0\,. \tag{4.2.20}$$

The definition of action (4.2.5) yields

$$I = \frac{1}{2n_0}\omega_0\rho^2 \approx \frac{1}{2n_0}\omega_0\rho_0^2 + \frac{\omega_0\rho_0}{n_0}\tilde{\rho}\,. \qquad (4.2.21)$$

The change in the variable

$$\tilde{I} = I - \frac{1}{2n_0}\omega_0\rho_0^2 = \frac{\omega_0\rho_0}{n_0}\tilde{\rho} \qquad (4.2.22)$$

is canonical (shift by a constant). Within the same approximation and, as in the case of the Hamiltonian $\tilde{H}$, the pair of variables $(\tilde{\rho}, \theta)$ is canonical. Thus, the Hamiltonian equations of motion have the following form:

$$\dot{\tilde{\rho}} = -\frac{n_0}{\rho_0\omega_0}\frac{\partial \tilde{H}_0}{\partial \theta}\,, \quad \dot{\theta} = \frac{n_0}{\rho_0\omega_0}\frac{\partial \tilde{H}_0}{\partial \tilde{\rho}}\,. \qquad (4.2.23)$$

In fact, they coincide with those derived from (4.2.7) under the condition (4.2.20).

After denoting

$$\omega_w = -\sigma\left(\frac{2}{\pi}\right)^{1/2}\frac{\epsilon n_0\omega_0}{(k\rho_0)^{3/2}} \qquad (4.2.24)$$

and defining a new Hamiltonian,

$$H_w = \omega_w \cos\xi \cos\theta\,, \qquad (4.2.25)$$

where $\xi = k\tilde{\rho}$, the equations

$$\dot{\xi} = \frac{\partial H_w}{\partial \theta}\,, \quad \dot{\theta} = -\frac{\partial H_w}{\partial \xi} \qquad (4.2.26)$$

are equivalent to (4.2.23). The Hamiltonian $H_w$ can be called the Hamiltonian of the web-tori [see footnote (1)] and $\omega_w$ the frequency of the small oscillations.

Rewriting (4.2.26) in an explicit form as

$$\dot{\xi} = -\omega_w \cos\xi \sin\theta\,, \quad \dot{\theta} = \omega_w \sin\xi \cos\theta \qquad (4.2.27)$$

and using (4.2.25) and integrating (4.2.27), one obtains

$$\sin\theta = \kappa\,\mathrm{sn}(\omega_w(t - t_0); \kappa)\,, \qquad (4.2.28)$$

where sn is the Jacobian elliptic function, $t_0$ is the time instant at which $\theta = 0$, and $\kappa$ is the modulus of the elliptic function

$$\kappa = (1 - H_w^2/\omega_w^2)^{1/2} . \tag{4.2.29}$$

Using solution (4.2.28) and expressions (4.2.25) and (4.2.29), one derives

$$\sin\xi = \kappa \frac{\mathrm{cn}(\omega_w(t-t_0);\kappa)}{\mathrm{dn}(\omega_w(t-t_0);\kappa)} . \tag{4.2.30}$$

Both solutions (4.2.28) and (4.2.30) are periodic functions of time. The period of nonlinear oscillations is

$$T(H_w) = \frac{1}{|\omega_w|} 4K(\kappa) , \tag{4.2.31}$$

where $K(\kappa)$ is a full elliptic integral of the first kind. When $\kappa \to 0$,

$$T_w = 2\pi/|\omega_w| , \tag{4.2.32}$$

that is, the frequency of small oscillations coincides with frequency (4.2.24). Near the separatrix, $\kappa \to 0$ and it follows from Eq. (4.2.31) that

$$\begin{aligned} T(H_w) &= \frac{4}{|\omega_w|} \ln \frac{4}{(1-\kappa^2)^{1/2}} = \frac{4}{|\omega_w|} \ln \frac{4|\omega_w|}{|H_w|} \\ &= \left(\frac{\pi}{2}\right)^{1/2} \frac{4}{\epsilon n_0 \omega_0} (k\rho_0)^{3/2} \ln \left[ 4\epsilon \frac{\omega_0^2}{k^2} \left(\frac{2}{\pi k \rho_0}\right)^{1/2} \frac{1}{|\tilde{H}_0|} \right] . \end{aligned} \tag{4.2.33}$$

On the separatrices, $H_w = 0$. From (4.2.25), it follows that the four separatrices in Fig. 4.2.2 are defined by the equations

$$\cos\xi = 0 , \quad \xi = \pm\pi/2 \tag{4.2.34}$$

and

$$\cos\theta = 0 , \quad \theta = \pm\pi/2 . \tag{4.2.35}$$

They correspond to the four sides of the square (under the approximation $\tilde{\rho} \ll \rho_0$) in Fig. 4.2.2. The solution for (4.2.34),

$$\sin\theta = \pm\tanh[2|\omega_w|(t-t_0)] , \quad \xi = \pm\pi/2 , \tag{4.2.36}$$

corresponds to two horizontal separatrices and for (4.2.35), to two vertical ones:

$$\sin\theta = \mp\tanh[2|\omega_w|(t-t_0)]\,, \quad \xi = \mp\pi/2\,. \tag{4.2.37}$$

The closed trajectories defined by the Hamiltonians (4.2.19) and (4.2.25) become the cross-sections of the invariant tori if one complements the phase space $(I,\theta)$ with the "time" variable [see also footnote (1)] in the usual way (that is, by taking into account the perturbation $\tilde{V}$ which is periodic in time). The invariant tori lying inside a web is referred to as web-tori and they differ from the KAM-tori in that their period $T_w$ is dependent on $\epsilon$. In the case of KAM-tori, $T \sim 1/\epsilon^{1/2}$ while for the web-tori, $T_w \sim 1/\epsilon$.

Another important difference between KAM-tori and web-tori is the angle at which their separatrices cross. It can be arbitrarily small for KAM-tori and for the web-tori, it is a constant [$\pi/2$ in the case of (4.2.25)], depending on the web's structure.

One can state that in the case of the web-tori, the degeneracy condition

$$\frac{d^2H_0(I)}{dI^2} = 0 \tag{4.2.38}$$

allows radial infinite diffusion while condition (4.1.4) does not allow it to take place in the KAM-case and 1 1/2 degrees of freedom.

## 4.3 Width of the Stochastic Web

For the perturbed oscillator model (4.1.2), the Hamiltonian was rewritten as (4.2.11) which transforms into

$$H_{n_0}(\tilde{I},\theta) = \epsilon\frac{\omega_0^2}{k^2}J_{n_0}(k\tilde{\rho})\cos\theta + \epsilon\frac{\omega_0^2}{k^2}\sum_{m\neq n_0} J_m(k\tilde{\rho})\cos\left[\frac{m}{n_0}\theta n - \left(1-\frac{m}{n_0}\right)\nu t\right] \tag{4.3.1}$$

for the resonance (4.1.5) with

$$\tilde{\rho} = (2n_0\tilde{I}/\omega_0)^{1/2} \tag{4.3.2}$$

[see (4.2.9)]. Unperturbed motion is defined by the first term in (4.3.1) which corresponds to the Hamiltonian (4.2.12) (it was considered in Section 4.2). Its phase plane is covered with a web-like net of separatrices (Fig. 4.2.1). The second term in (4.3.1) defines a perturbation,

$$V_{n_0} = \epsilon \frac{\omega_0^2}{k^2} \sum_{m \neq n_0} J_m[(k(2n_0 \tilde{I}/\omega_0)^{1/2}] \cos\left[\frac{m}{n_0}\theta - \left(1 - \frac{m}{n_0}\right)\nu t\right] \quad (4.3.3)$$

whose presence destroys the separatrices and replaces them with stochastic layers. The width of the stochastic layers can be obtained using the same scheme described in Chapter 2.

To simplify the estimation, the inequality (4.2.16), that is,

$$k\rho_0 = k(2n_0\tilde{I}/\omega_0)^{1/2} \gg 1 \quad (4.3.4)$$

and the two terms with $m = n_0 \pm 1$ in the sum (4.3.3) are retained. It yields

$$V_{n_0} \approx 2\epsilon \frac{\omega_0^2}{k^2} \left(\frac{2}{\pi k \rho_0}\right)^{1/2} \sigma \sin k\tilde{\rho} \sin\theta \sin\left[\frac{m}{n_0}(\theta + \nu t)\right] \quad (4.3.5)$$

for the perturbation (4.3.3) where the sign function, $\sigma$, is similar to that found in (4.2.19) and indicates what type of cells is being considered. Thus, the entire problem is described by the following Hamiltonian:

$$H_{n_0} = \sigma\epsilon \frac{\omega_0^2}{k^2} \left(\frac{2}{\pi k \rho_0}\right)^{1/2} \left\{-\cos k\tilde{\rho}\cos\theta + 2\sin k\tilde{\rho}\sin\theta\sin\left[\frac{1}{n_0}(\theta + \nu t)\right]\right\}. \quad (4.3.6)$$

The calculations, similar to those found in Chapter 2, are omitted and the final results are written down. The width of the separatrix splitting, or Melnikov integral [see (2.3.5) and (2.3.20)], is

$$\Delta H_{n_0} = 2\pi^2 \frac{\omega_0^2}{k^2} \exp\left\{-\frac{1}{\epsilon}\left(\frac{\pi}{2}\right)^{5/2} (k\rho_0)^{1/2}\right\} \sin\phi_0 , \quad (4.3.7)$$

where $\phi_0$ is a phase that depends on the separatrix being considered and on an initial time instant. The corresponding width of the stochastic web is

$$\Delta H_s = 2^{1/2}\pi^{7/2}\frac{1}{\epsilon}(k\rho_0)^{1/2}\frac{\omega_0^2}{k^2}\exp\left\{-\frac{1}{\epsilon}\left(\frac{\pi}{2}\right)^{5/2}(k\rho_0)^{1/2}\right\}. \tag{4.3.8}$$

Expression (4.3.8) indicates that the width of the web decreases quickly as the radius $\rho_0$ grows, that is, the width of the web for fairly distant cells is exponentially small.[2]

## 4.4 Transition from the KAM-Tori to Web-Tori

One now turns to a comparison of the equations for the perturbed pendulum and oscillator, (4.1.10) and (4.1.11), respectively, by using the expansion (4.1.13). The Hamiltonian (4.2.11) for the perturbed oscillator can be rewritten as:

$$H = (n_0\omega_0 - \nu)I - \frac{n_0^2}{6}I^2\sin^4\phi + \epsilon\frac{\omega_0^2}{k^2}J_{n_0}(k\rho)\cos\theta + \epsilon\frac{\omega_0^2}{k^2}\sum_{m\neq n_0}J_m(k\rho)$$

$$\cdot\cos\left[\frac{m}{n_0}\theta - \left(1 - \frac{m}{n_0}\right)\nu t\right]. \tag{4.4.1}$$

For the purpose of estimation:

1. Consider only the part which includes the first three terms and keep as a perturbation only the term with $m = n_0 \pm 1$ in the sum.
2. Ignore the non-resonant detuning arising from the first term in $H$, that is, put $n\omega_0 = \nu$.
3. Replace $\sin^4\phi$ by its average value 3/8.

Elliptic and hyperbolic points of the unperturbed Hamiltonian satisfy Eqs. (4.2.13), (4.2.14) and (4.2.15). Non-trivial solutions exist for these equations if

$$\frac{n_0^2}{8}I < \epsilon\frac{\omega_0^2}{k^2}\frac{dJ_{n_0}(k\rho)}{dI}. \tag{4.4.2}$$

---

[2]The width, $\Delta H_s$, of the stochastic web was obtained in [FMIT 77]. See also [ZSUC 91].

Without sacrificing generality, when $\omega_0 = 1$ and the "effective" perturbation

$$\bar{\epsilon} = \epsilon\omega_0^2/k^2 \tag{4.4.3}$$

is introduced, condition (4.4.2) becomes

$$\bar{\epsilon}k^4 \cdot 6J'_{n_0}(\xi)/\xi^3 > 1\,, \tag{4.4.4}$$

where $\xi = k\rho$ is a dimensionless parameter required for the determination of the stationary point radius. The criterion for its existence follows immedately from (4.4.4):

$$\bar{\epsilon}k^4 > \xi^3/6J'_{n_0}(\xi)\,, \tag{4.4.5}$$

where $\xi$ is proportional to $I^{1/2}$ and characterises a level of nonlinearity, that is, expression (4.4.5) possesses perturbation on its left-hand side and nonlinearity on its right-hand side.

If condition (4.4.5) is valid, the solutions in (4.4.4) will then exist. The stochastic web can also be implemented in the separatrix loop of the perturbed pendulum. Numerical examples on this phenomena are presented in Fig. 4.4.1. Figure 4.4.1(a) shows the separatrix loop of the perturbed pendulum with a thin stochastic layer instead of a destroyed separatrix. At the same time, fragments of the web-tori with stochastic web between them, instead of the regular KAM-tori, were implanted around the central elliptic point. The corresponding web-structure is magnified in Figs. 4.4.1(b) and (c).

It follows from (4.4.5) that for an appropriate vicinity in the resonance condition and arbitrarily small perturbation $\bar{\epsilon} \to 0$, it is possible to derive $k$ such that the KAM-tori structure does not exist in the vicinity of the origin.

A deviation from the resonance $\delta\omega = \nu - n\omega_0 \neq 0$ can be considered in the same way as small nonlinearity.[3]

[3]The described web-structure and transition between web-tori and KAM-tori follow [CNPSZ 87, 88]. The phenomenon of radial transport arising from the stochastic web has various applications in plasma physics [LiL 92, HI 96].

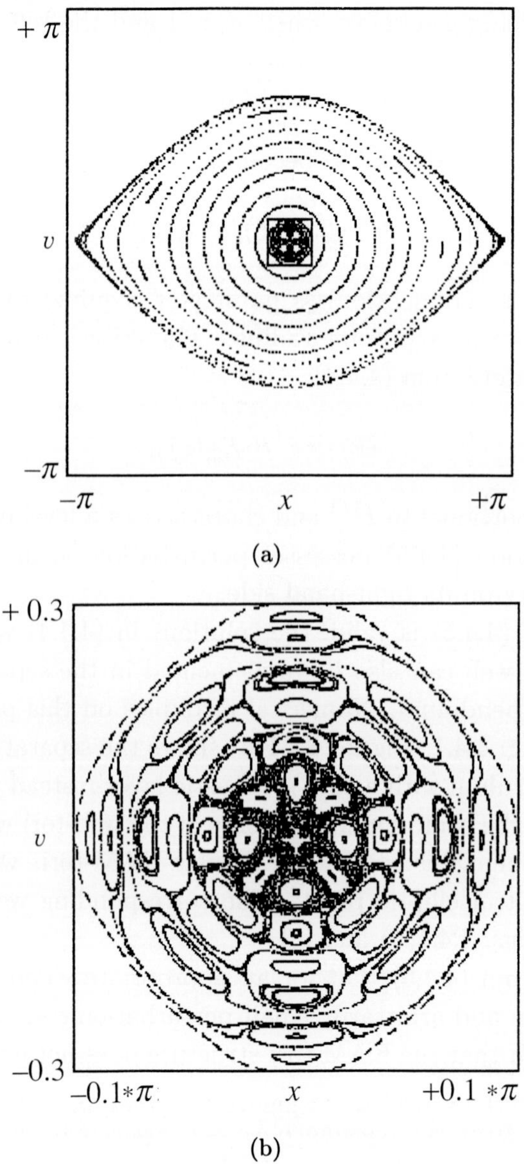


Fig. 4.4.1. Implantation of part of the web in the separatrix loop of the perturbed pendulum problem with $\omega_0 = 1$, $\nu = 4$, $k = 75$ and $\epsilon/k = 3/200$: (a) separatrix loop with implemented web; (b) magnification of the central part of (a); (c) the same as (b) but with $k = 300$ and $\epsilon/k = 1/400$.

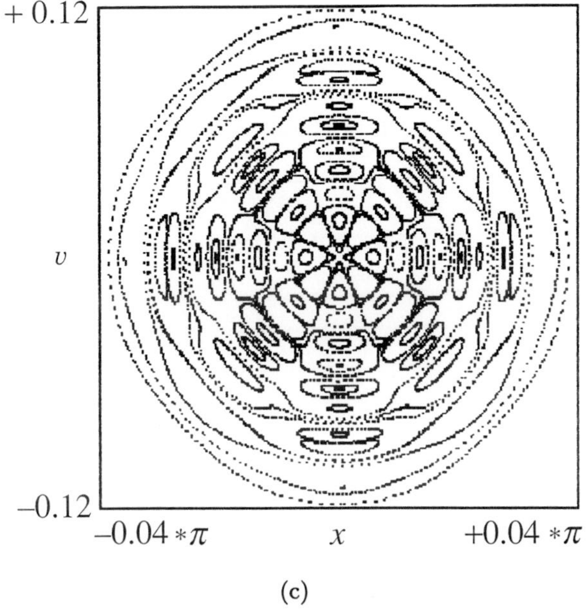


(c)

Fig. 4.4.1. (*Continued*)

## Conclusions

1. A typical physical situation arises when the degeneracy condition

$$\left|\frac{\partial^2 H_0}{\partial I_k \partial I_\ell}\right| = 0$$

occurs for a non-perturbed Hamiltonian, $H_0 = H_0(I_1, \ldots, I_N)$, and the KAM theory cannot be used (at least with immediate effect). In this case, infinite stochastic webs can even appear for 1 1/2 degrees of freedom. The corresponding invariant web-tori differ from those of the KAM. The level of degeneracy corresponds to the level of nonlinearity, $\alpha_{n\ell}$, which for 1 1/2 degrees of freedom is simply

$$\alpha_{n\ell} = \frac{I}{\omega_0}\left|\frac{d\omega_0}{dI}\right| = \frac{I}{\omega_0}\left|\frac{d^2 H_0}{dI^2}\right| .$$

2. For the finite values of the nonlinear parameter, $\alpha_{n\ell}$, and perturbation parameter, $\epsilon$, the phase space pattern and properties of chaotic dynamics depend on the relationship between $\alpha_{n\ell}$ and $\epsilon$. This is in spite of the fact that both of them are fairly small.

# *Chapter 5*

# FRACTALS AND CHAOS

## 5.1 Fractal Dynamics

The structures generated by chaotic dynamics in the phase space are so complex that they cannot be described in a conventional way. One is therefore reduced to dealing with objects that are unusual, "wild" and "strange". One can use the notion of fractal objects in chaotic dynamics but, as will be seen below, even this proves insufficient in describing chaos.

The main feature of a fractal object is its self-similarity. A discussion of the strict definition for fractals, which probably does not exist, is omitted here and one shall dwell upon the specific properties of a set of elements $\{u_j\}$ distributed in the phase space. One can imagine a set of points on the Poincaré map in the phase space. $\{u_j\}$ is then a set of functions for the co-ordinates $r_j$ in the map with the subscript $j$ indicating the iteration number of the map. There are many different ways to characterise the distribution of the points on the map in the phase space. The normal physical approach is to either introduce a distribution function, $F(r_j, p_j; t)$, or to measure them in the phase space. This measure should be smoothed over or coarse-grained in some way to simplify the description of the dynamics by reducing the information on the trajectories. Nevertheless, even a typical coarse-graining of the distribution function does not remove some underlying "wild" structures of the space, which are a support for chaotic trajectories. A brief glance at Fig. 3.1.1 warns that the object of consideration, the phase space of chaos, is highly complex and can be analysed, in part, via its fractal properties.[1]

One begins with some simple definitions. Consider a set of elements, say points, embedded in an arbitrary dimension space confining these elements. Cover the elements by using a set of balls with the same diameter $\epsilon$ and consider only minimal sets of balls, $\{u_j(\epsilon)\}$, where $u_j(\epsilon)$ refers to a ball with the diameter $\epsilon$ and $j$ is its number. After letting $N_\epsilon$ be the number of balls in the minimal set, there are two ways to proceed:

(i) **Box dimension.** Consider the limit

$$d_C = \lim_{\epsilon \to 0} \frac{\ln N_\epsilon}{\ln(1/\epsilon)} . \tag{5.1.1}$$

$d_C$ is called a box dimension because of the relation

$$\epsilon^{d_C} \cdot N_\epsilon \sim \text{const} \tag{5.1.2}$$

in the limit $\epsilon \to 0$, that is, $d_C$ is an "effective" dimension of small boxes that cover the set of points.

(ii) **Hausdorff dimension.** Consider the sum

$$\mathbf{S} = \lim_{\epsilon \to 0} \sum_j \epsilon_j^d , \tag{5.1.3}$$

where $j$ labels a ball and the summation is performed over a minimal set of balls with diameter $\epsilon_j$, which is assumed to be in the interval $\epsilon_j \in (\epsilon, \epsilon + \delta\epsilon)$. For $\epsilon \to 0$, the number of terms in (5.1.3) tends toward infinity and the result is

$$\mathbf{S} = \begin{cases} \infty , & d < d_h \\ \text{const.} > 0 , & d = d_H , \\ 0 , & d > d_h \end{cases} \tag{5.1.4}$$

[1]For more information on fractals, the reader is referred to the book by B. Mandelbrot [Ma 82]. For a list of rigorous definitions on topics related to fractals and fractal dimensions, the reader can refer to the article by Y. Pesin [Pe 88].

where $d_H$ is some number called the Hausdorff dimension. If all the $\epsilon_j$ are the same, expression (5.1.3) yields

$$\mathbf{S} = \lim_{\epsilon \to 0} N(\epsilon)\epsilon^{d_H} \tag{5.1.5}$$

and $d_C$ will coincide with $d_H$. In fact, there are some exceptions which will not be discussed here.

An example of a fractal object is shown in Fig. 5.1.1 where the proliferation of balls will continue till it reaches infinity. This object is called the Serpinsky carpet. When

$$S_{n+1} = \lambda_S S_n \,, \tag{5.1.6}$$

Fig. 5.1.1. Serpinsky carpet as a fractal object.

where $S_n$ is the area in a circle of the $n$-th generation and $\lambda_S$ is a constant, each generation gives birth to $q$ circles ($q = 8$ in Fig. 5.1.1). Thus, for the $n$-th generation,

$$N_n = q^n \,, \quad S_n = \lambda_S^n \tag{5.1.7}$$

and

$$d_H = \lim_{n\to\infty} \frac{\ln q^n}{|\ln \lambda_S^{n/2}|} = \frac{\ln q}{|\ln \lambda_S^{1/2}|} \,. \tag{5.1.8}$$

Since $q = 8$ and $\lambda_S \geq 1/9$, one has $d_H \leq \ln 8/\ln 3 < 2$.

## 5.2 Generalised Fractal Dimension

As will be seen later, the different scaling properties of dynamical systems can be formulated in precisely the same way as (5.1.4). However, they do not have a specific relation with dimensions. Such a situation can be described by a dimension-like characteristic introduced by Ya. Pesin (see footnote [1]). Here it is called a generalised dimension.

A partition, $\{u_j\}$, is made in the space by using the element $u_j$, and $\xi(u_j)$ and $\eta(u_j)$ are some "good" functions defined on the partition element $u_j$. One can select the elements of the partition in any convenient shapes such that its effective diameter satisfies the condition diam $u_j \leq \epsilon$. Consider a limit of the minimal sum,

$$\mathbf{S} = \lim_{\epsilon\to 0} \sum_j \xi(u_j)\eta^d(u_j) \,, \tag{5.2.1}$$

where the minimal sum refers to the attempt to cover the set of points with co-ordinates $r_j$ by using a minimum number of the partition element $u_j = u_j(r, \epsilon)$. Hence a constant $d_g$ similar to $d_H$ in (5.1.4), for which the sum $0 < \mathbf{S} < \infty$, exists. The number $d_g$ is the Pesin generalised dimension. The following examples illustrate the use of $d_g$ in different situations:

1. $\xi = 1$, $\eta(u_j) = \epsilon_j$. This yields the Hausdorff dimension [see (5.1.3)].
2. $\xi = 1$, $\eta(u_j) = \eta(\epsilon_j)$. This yields the so-called Carathéodory dimension.

3. The example that follows is more sophisticated. First, $\xi = 1$ and a large integer, $n$, is introduced. A small number, $\epsilon$, and a point $x$ in the phase space are then introduced. The image of the point $x$ along the trajectory after time interval $T$ is also denoted as $\hat{T}_x$, that is, $\hat{T}$ is simply a time-shift (Liouvillian) operator.

Finally, one considers all the points $y$ for which the distance $\rho$ satisfies the condition of a finite points separation during time $nT$:

$$\rho(\hat{T}^n x, \hat{T}^n y) = \rho(x(t_0 + nT), y(t_0 + nT)) \leq \epsilon\,. \tag{5.2.2}$$

After stipulating that

$$\eta(y_j) = e^{-n}\,, \tag{5.2.3}$$

where the points $y_j$ satisfy the condition (5.2.2), the non-trivial part of expression (5.2.3) is seen in the fact that although $\eta$ is a constant, the summation in (5.2.1) should be performed over all points $y_j$ which do not have a trivial definition. After substituting (5.2.3) in (5.2.1), one derives either

$$\mathbf{S} = \text{const} \sim \lim_{\epsilon \to 0} N_\epsilon(n) e^{-nd_{KS}} \tag{5.2.4}$$

or

$$N_\epsilon(n) \sim e^{nd_{KS}}\,, \tag{5.2.5}$$

where $N_\epsilon(n)$ is the number of points $y_j$ separated from $x$ on a distance $\epsilon$ during the dimensionless time $n$. A dimension-like characteristic, $d_{KS} > 0$, which provides the condition (5.2.5) for the exponential growth in the number of partition elements $u_j$, is the Kolmogorov–Sinai entropy:

$$d_{KS} = h_{KS}\,. \tag{5.2.6}$$

Indeed, the condition (5.2.5) corresponds to the exponential growth of the enveloped volume or (which is the same) the number of points with their distance bounded by $\epsilon$ from the central one.

Definition (5.2.1) and the existence of the limit provide a broad view of the notion of fractal dimension which is not only related to the geometrical characteristic of an object. An example in the next section confirms the truth of this statement.

## 5.3 Renormalisation Group and Generalised Fractal Dimension

Consider an object, $\mathcal{A}_0$, and its subsequent transformations:

$$\mathcal{A}_1 = \hat{T}\mathcal{A}_0\,, \quad \mathcal{A}_n = \hat{T}\mathcal{A}_{n-1} = \cdots = \hat{T}^n\mathcal{A}_0\,. \tag{5.3.1}$$

For example, $\mathcal{A}_j$ may be a domain in the phase space and $\hat{T}$ is either a time-shift operator at interval $T$ or any other transformation in the domain $\mathcal{A}_j$. Let $\mathcal{A}_j = \mathcal{A}_j(r)$ and $r$ be a characteristic diameter of $\mathcal{A}_j$. Consider also the transformation of $r$ by a factor $\lambda$:

$$\hat{S}\mathcal{A}_j(r) = \mathcal{A}_j(\lambda r)\,. \tag{5.3.2}$$

After introducing a function, $V(\mathcal{A}_n)$, defined for the domains $\mathcal{A}_j$ and assuming, for example, that $V(\mathcal{A}_n)$ is the volume of $\mathcal{A}_n$, one then considers the transform

$$\hat{R} = \hat{S}\hat{T} \tag{5.3.3}$$

after assuming that the existence of self-similarity for volume $V$ is

$$V(\hat{R}\mathcal{A}_n) = \lambda^{d_R} V(\mathcal{A}_n)\,, \tag{5.3.4}$$

where $d_R$ is an exponent that characterises the self-similarity of $\{\mathcal{A}_j\}$ under the renormalisation transform $\hat{R}$. In general, condition (5.3.4) is only valid for the limit

$$\begin{aligned} d_R &= \lim_{n\to\infty} \frac{\ln[V(\hat{S}\mathcal{A}_{n+1})/[V(\mathcal{A}_n)]}{\ln\lambda} \\ &= \lim_{n\to\infty} \frac{\ln[\hat{V}(\hat{R}\mathcal{A}_n)/V(\mathcal{A}_n)]}{\ln\lambda}\,, \end{aligned} \tag{5.3.5}$$

where the definitions (5.3.1) to (5.3.3) have been used. If the sequence $\{\mathcal{A}_0, \mathcal{A}_1, \ldots\}$ has a fixed point $\mathcal{A}^*$, it follows from (5.3.5) that

$$d_R = \ln[V(\hat{S}\mathcal{A}^*)/V(\mathcal{A}^*)]/\ln\lambda\,, \tag{5.3.6}$$

or the equivalent equation

$$V(\hat{S}\mathcal{A}^*) = \lambda^{d_R} V(\mathcal{A}^*)\,, \tag{5.3.7}$$

can be used to define $\mathcal{A}^*$ and $d_R$.

One can see that the definition of $d_R$ coincides with that of $d_C$ in (5.1.1). Indeed, as a result of (5.3.2),

$$V(\hat{S}\mathcal{A}^*(r)) = V(\mathcal{A}^*(\lambda r)) \tag{5.3.8}$$

and one should take into account the fact that $V(\mathcal{A}^*)$ in (5.3.6) does not depend on $\lambda$ while $V[\hat{S}\mathcal{A}^*(r)]$ does. One thus derives from (5.3.6)

$$d_R = \lim_{\lambda\to\infty} \frac{\ln \hat{V}(\mathcal{A}^*(\lambda r))}{\ln \lambda} = d_C\,. \tag{5.3.9}$$

If $r$ is not a diameter and $V$ is not the volume, the definition of (5.3.6) for $d_R$ then has the same structure as (5.2.1). It is written as

$$\mathbf{S} \sim N(u_j)\cdot \eta^d(u_j) \sim \text{const.} \tag{5.3.10}$$

The condition for the existence of a limit when $N \to \infty$, $\eta \to 0$ yields

$$d = \lim_{N\to\infty} \frac{\ln N(u_j)}{\ln(1/\eta(u_j))}\,, \tag{5.3.11}$$

which is the same formula for (5.3.9). The crucial reason for considering $d_R$ as a particular case of $d_g$ is due to the existence of both the Renormalisation Group and the fixed point.

## 5.4 Multi-Fractal Spectra

In this section, one describes a very strong generalisation of the notion of fractals. A situation may occur when the set of objects, $\{u_j(r_j)\}$ and stuck to points with co-ordinates $r_j$, cannot be described using only the self-similarity exponent, $d_g$, and more of such exponents is therefore needed to do so. One tries to be more specific and again partitions the space by using a set of elements, $\{u_j\}$, and the same condition mentioned in Section 5.1:

$$\ell_j \equiv \operatorname{diam} u_j \le \epsilon\,. \tag{5.4.1}$$

Assuming that for a fairly small $\epsilon$, the probability of finding a particle in the cell $u_j$ is

$$P_j = \text{const} \cdot \ell_j^{\gamma_j} , \tag{5.4.2}$$

where $\gamma_j$ is a characteristic exponent defined locally for the $j$-th cell, a generating sum is constructed:

$$Z_\ell(q) = \sum_j P_j^q = \text{const} \sum_j \ell_j^{\gamma_j q} . \tag{5.4.3}$$

On the right-hand side, the summation over $j$ runs throughout the entire set $\{u_j\}$. The aim is to rewrite the sum (5.4.3) in a way that takes into account the existence of different cells with the same value $\gamma$. When $\Delta N(\gamma)$ is the number of cells with the same value $\gamma$, instead of (5.4.3), one can then stipulate that

$$Z_\ell(q) = \text{const} \sum_\gamma \Delta N(\gamma) \ell^{q\gamma} , \tag{5.4.4}$$

where a desired change of the variables from $j$ to $\gamma$ is performed. The same condition in (5.4.2) can be applied to $\Delta N(\gamma)$, that is,

$$\Delta N(\gamma) = \ell^{-f(\gamma)} , \tag{5.4.5}$$

where a new important exponent, $f(\gamma)$, is introduced. It is called a dimension spectral function.[2]

The calculations that follow are very formal. The sum of (5.4.4) and (5.4.5) is

$$Z_\ell(q) = \text{const} \sum_\gamma \ell^{q\gamma - f(\gamma)} . \tag{5.4.6}$$

There are two ways to proceed. First, one can apply the same consideration in (5.2.1) where the generalised dimension was introduced. Second, which is more typical in physical consideration, the sum (5.4.6) is replaced by integration. The expression

$$\Delta N(\gamma) \to dN(\gamma) = \rho(\gamma) \ell^{-f(\gamma)} d\gamma \tag{5.4.7}$$

[2] Fractal and multi-fractal analysis of dynamical systems with chaotic motion became a routine method after the publication of a series of pioneering works: [HP 83], [HP 84], [FP 85], [JKLPS 85] and [HJKPS 86]. The rigorous consideration provided in [Pe 88] confirmed the basic physical concepts. For a review of the different topics related to fractals and multi-fractals, see [PV 87].

is used instead of (5.4.5). Here $\rho(\gamma)$ is a density function which does not depend on $\ell$ or if it is, it does so slowly. Hence

$$Z_\ell(q) = \int d\gamma\, \rho(\gamma) \exp\left\{-\left(\ln\frac{1}{\ell}\right)[q\gamma - f(\gamma)]\right\}, \tag{5.4.8}$$

where the const. in (5.4.6) is included in $\rho(\gamma)$. Consider the limit

$$Z(q) = \lim_{\ell\to\infty} Z_\ell(q)\,. \tag{5.4.9}$$

Without any special comments, one can state that

$$n \equiv \ln\frac{1}{\ell} \tag{5.4.10}$$

and consider a limit $n \to \infty$ instead of $\ell \to 0$. After applying the saddle-point method, one arrives at

$$Z(q) \sim \exp\{-n[q\gamma_0 - f(\gamma_0)]\}\,, \quad (n\to\infty) \tag{5.4.11}$$

where the saddle-point, $\gamma_0$, satisfies the equation

$$\frac{df(\gamma_0)}{d\gamma_0} = q\,. \tag{5.4.12}$$

The final expression, (5.4.11), can also be rewritten as

$$Z(q) = \exp(-n\mathcal{D}(q))\,, \quad (n\to\infty) \tag{5.4.13}$$

where

$$\mathcal{D}(q) = \gamma_0 q - f(\gamma_0)\,. \tag{5.4.14}$$

The new variable, $\mathcal{D}(q)$, is a new dimension type characteristic of the system. $\mathcal{D}(q)$ and $f(\gamma_0)$ form a Legendre transform pair.[3]

It is also convenient to introduce the so-called Rényi dimension $D_q$,

$$\mathcal{D}(q) = (q-1)D_q\,, \tag{5.4.15}$$

[3]Rényi's dimension was introduced in [Re 70].

and consider its meaning for different values of $q$. For $q = 0$, one derives

$$D_q = -\mathcal{D}(q) \tag{5.4.16}$$

and (5.4.13) yields

$$Z(0) = e^{nD_0} = 1/\ell^{D_0} = N(\ell)\,, \quad (n \to \infty,\, \ell \to 0) \tag{5.4.17}$$

which corresponds to (5.1.2). Hence $D_0$ is similar to $d_C$, that is, the box dimension.

To obtain $D_1$, one considers the definition (5.4.3),

$$Z_\ell(1+\delta_q) = \sum_j P_j^{1+\delta_q} = 1 + \delta_q \sum_j P_j \ln P_j\,, \tag{5.4.18}$$

where $\delta_q = q - 1 \ll 1$. From (5.4.13) and (5.4.15), one arrives at

$$Z_\ell(1+\delta_q) = \ell^{\delta_q \cdot D_q}\,. \tag{5.4.19}$$

A comparison of (5.4.18) and (5.4.19) yields

$$D_1 = \lim_{\ell\to 0}[\ln \sum_j P_j \ln P_j]/\ln(1/\ell) = \lim_{\ell \to 0} \frac{\ln N(\ell)}{\ln 1/\ell}\,, \tag{5.4.20}$$

where $P_j = 1/N(\ell)$ with the number of cells $N(\ell)$ is used. [Compare this to (5.1.1) and (5.1.2)]. The magnitude

$$S_{\text{inf}}(\ell) = \sum_j P_j \ln P_j = \ln N(\ell) \tag{5.4.21}$$

is also known as information entropy. It follows from (5.4.20) and (5.4.10) that

$$N(\ell) \sim \ell^{-D_1}\,, \tag{5.4.22}$$

and in correspondence to (5.4.17),

$$D_1 = D_0\,. \tag{5.4.23}$$

One more dimension, $D_2$, is of especial interest since it defines the Grassberger-Hentschel-Procaccia correlation dimension.[4] Consider the pair correlation function between points $r_j$ and $r_k$:

$$C_{\text{cor}}(\ell) = \frac{1}{N_\ell^2} \sum_{j,k} \theta(\ell - |r_j - r_k|) \tag{5.4.24}$$

where

$$\theta(\xi) = \begin{cases} 1, & \xi \geq 0 \\ 0, & \xi < 0 \end{cases}.$$

In the sum (5.4.24), only 1 is obtained when $r_j$ and $r_k$ belong to the same cell. If $P_j$ is the probability for a particle to exist in the $j$-th cell, then the probability of two particles existing in the same cell is simply $P_j^2$ and

$$C_{\text{cor}} = \sum_j P_j^2 = Z_\ell(2) \sim \ell^{D_2} \tag{5.4.25}$$

based on definition (5.4.3). Hence

$$D_2 = \lim_{\ell \to 0} \frac{\ln C_{\text{cor}}(\ell)}{|\ln \ell|}. \tag{5.4.26}$$

For the limit cases $q = \pm\infty$, one derives from definition (5.4.3) the following:

$$\begin{aligned} Z_\ell(q \to \infty) &\sim P_{\max}^q \sim \ell^{\gamma_{\max} q} \sim \ell^{q D_\infty} \\ Z_\ell(q \to -\infty) &\sim P_{\min}^q \sim \ell^{\gamma_{\max} q} \sim \ell^{q D_{-\infty}} \end{aligned}. \tag{5.4.27}$$

Therefore,

$$D_\infty = \gamma_{\min}, \quad D_{-\infty} = \gamma_{\max}. \tag{5.4.28}$$

A typical behaviour of the dimension $D_q$ as a function of $q$ and, correspondingly, of the spectral function $f(\gamma)$ is shown in Fig. 5.4.1. In the following chapter, a generalisation of the multi-fractal notions with regards to the time events of dynamical systems is presented.

---

[4]See [GHP 83] for the original publication and [PW 97] for the discussions and rigorous results. The correlation dimension is very convenient for different practical estimations.

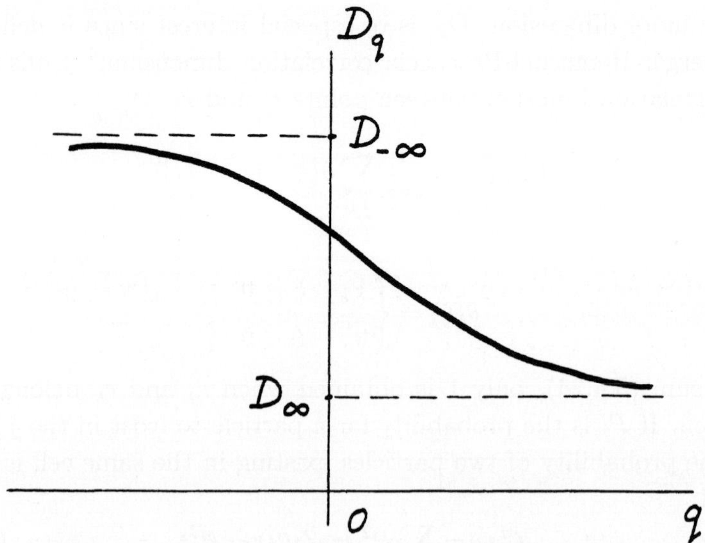


Fig. 5.4.1. A typical behaviour of the fractal dimension $D_q$.

## Conclusions

1. Chaotic trajectories represent a typical fractal or multi-fractal object. The simplest definitions, such as the Hausdorff or box dimensions, are inadequate in describing a chaotic object. The generalised dimension $d$, introduced by Y. Pesin, provides a more adequate description of chaotic dynamics. The dimension $d$ is such that the minimal sum,

$$\mathbf{S} = \lim_{\epsilon \to \infty} \sum_j \xi(u_j)\eta^d(u_j)\,,$$

   is finite and non-zero and the functions $\xi(u)$ and $\eta(u)$ are properly chosen.

2. A powerful tool in fractal theory of dynamical systems is the multi-fractal spectral function, $f(\gamma)$, introduced by Hentchel and Procaccia.

3. The renormalisation group property of a system can be related to the system's fractal properties.

*Chapter 6*

# POINCARÉ RECURRENCES AND FRACTAL TIME

## 6.1 Poincaré Recurrences

When considering a Hamiltonian system that performs a finite motion in the phase space domain $\Gamma$, one uses a small area $A$ with phase volume $\Gamma_A < \Gamma$ and introduce a set of time instants, $\{t_j\}$, when the trajectory crosses the boundary of $A$ on the way from inside to outside. The intervals

$$\tau_j = t_{j+1} - t_j\,, \quad j = 0, 1, \ldots \tag{6.1.1}$$

are Poincaré cycles and $t_j$ are Poincaré recurrence times (see Fig. 6.1.1). Poincaré proved that for a finite and area-preserving motion, any

Fig. 6.1.1. Poincaré recurrence (cycle).

trajectory should return to an arbitrarily taken domain $A$ in a finite time and it should do so repeatedly for an infinite number of times. Exceptions exist only for zero measure orbits. This theorem has a long history in the problem of understanding the origin of statistical physics and thermodynamics.

Zermelo was the first to put the Poincaré recurrence theorem to sound use. In his critical comments on the Boltzmann kinetic theory, Zermelo assumed that recurrences occurred quasi-periodically. This appears to contradict Boltzmann's H-theorem on the entropy increase. Boltzmann had rightly argued that for a large number of particles, this time would be astronomically long. Unfortunately, his argument cannot be used for systems with dynamical chaos which can occur for as little as two interacting particles. For that reason, one is compelled to make a more thorough analysis of the problem of Poincaré recurrences.[1]

In fact, neither quasi-periodical recurrences nor any information on the nature of the $\{\tau_j\}$ sequence (6.1.1) follows from the Poincaré recurrences theorem. Under the conditions of dynamical chaos, sequence (6.1.1) is random and it is possible to raise the question of probability distribution for the recurrence time, $P_R(\tau)$, which has to be normalised as

$$\int_0^\infty P_R(\tau) d\tau = 1 . \tag{6.1.2}$$

To derive $P_R(\tau)$, one can introduce the distribution $P_R(\tau; \Delta\Gamma)$ of recurrences for a small domain $\Delta\Gamma$,

$$\frac{1}{\Delta\Gamma} \int_0^\infty P_R(\tau; \Delta\Gamma) d\tau = 1 , \tag{6.1.3}$$

and then consider a limit,

$$P_R(\tau) = \lim_{\Delta\Gamma \to 0} P_R(\tau; \Delta\Gamma)/\Delta\Gamma . \tag{6.1.4}$$

The importance of the existence of the limit (6.1.4) was proven by M. Kac which implies the presence of a finite mean recurrence time $\tau_R$, that is,

[1]For more discussion on the role of Poincaré recurrences in statistical physics, see [Za 85]. See [Ze 86] for the original publication by Zermelo.

$$\tau_R \equiv \int_0^\infty \tau P_R(\tau) d\tau < \infty \tag{6.1.5}$$

if the system's dynamics satisfy the conditions of compactness, ergodicity, and the existence of a non-zero measure $M(r) > 0$.[2] Although these conditions are sometimes easy to accept in different physical models, they cannot be applied *a priori* to real Hamiltonian systems with chaotic dynamics since the motion is definitely non-ergodic and the existence of positive measure is unclear thus far. Assuming that the properties (6.1.4) and (6.1.5) exist in Hamiltonian chaotic dynamics, it follows immediately follows from (6.1.5) that $P_R(\tau)$ has an asymptotic behaviour

$$P_R(\tau) \sim \tau^{-\gamma}, \quad (\tau \to \infty) \tag{6.1.6}$$

with the exponent

$$\gamma > 2\,. \tag{6.1.7}$$

For all known examples of simulation, condition (6.1.7) is valid although the motion is not ergodic.

## 6.2 Poissonian Distribution of Recurrences

In this section, a simplified view on the possible nature of the distribution function of recurrences for a system with fairly "good" chaotic properties is presented. As an example of such a good system, one can consider a two-dimensional map on the torus

$$\begin{pmatrix} x_{n+1} \\ y_{n+1} \end{pmatrix} \begin{pmatrix} K+1 & 1 \\ K & 1 \end{pmatrix} \begin{pmatrix} x_n \\ y_n \end{pmatrix}, \quad (x, y = \text{mod } 1) \tag{6.2.1}$$

where for the sake of simplicity, $K > 0$. It is area-preserving and has the constant Lyapunov exponents

[2] As mentioned by Boltzmann, the Poincaré recurrences theorem does not reveal anything about the properties of recurrence cycles. Recurrence cycles can be distributed in a random way without any contradictions in entropy growth. This is exactly what happens in systems with chaotic dynamics. However, dynamical chaos reveals several rich possibilities that one could not have possibly imagined not so long ago. They are discussed later in this chapter. The result for the finiteness of the mean recurrence time was obtained by M. Kac [Ka 58].

$$\lambda_{1,2} = 1 + K/2 \pm [(1 + K/2)^2 - 1]^{1/2}$$

or, for large values of $K$,

$$\lambda_1 = K\,, \quad \lambda_2 = 1/K\,. \tag{6.2.2}$$

The envelope phase volume in system (6.2.1) grows as

$$\bar{\Gamma}(t) \sim \exp h_{KS} t \tag{6.2.3}$$

with the Kolmogorov-Sinai entropy

$$h_{KS} = \ln K \tag{6.2.4}$$

in the case of (6.2.2) with large $K$.

This information is sufficient to obtain an estimation of the recurrences distribution $P_R(t)$ which should be proportional to the inverse phase volume filled by a trajectory during time $t$. This is in accordance with (6.2.3),

$$P_R(t) \sim \exp(-h_{KS} t) \tag{6.2.5}$$

or in the normalised form,

$$P_R(t) = h_{KS} \exp(-h_{KS} t) = \frac{1}{\tau_R} \exp(-t/\tau_R)\,, \tag{6.2.6}$$

with the mean recurrence time being

$$\tau_R = 1/h_{KS} = 1/\ln K\,, \tag{6.2.7}$$

where (6.2.4) is used in the model.

The Poissonian distribution for the first recurrences can be interpreted in another way. Consider a one-dimensional analog of (6.2.1),

$$x_{n+1} = \hat{T} x_n = K x_n\,, \quad (x, \text{mod } 1)\,, \quad K > 1 \tag{6.2.8}$$

where the period-$m$ orbit satisfies the condition

$$T^m x_n = x_n\,, \quad (x, \text{mod } 1)\,. \tag{6.2.9}$$

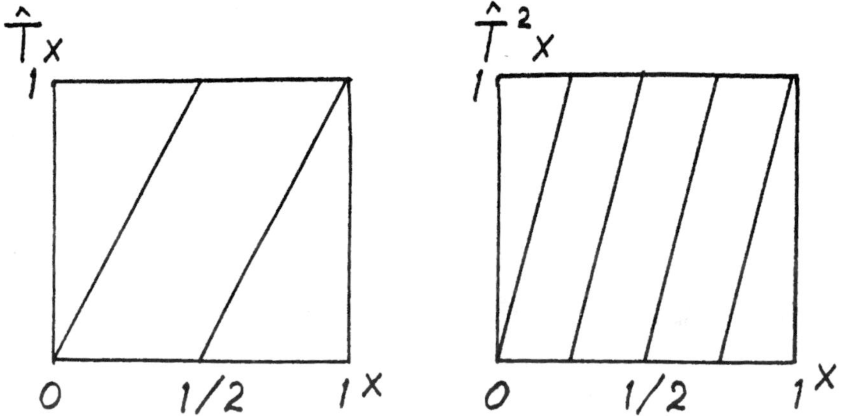


Fig. 6.2.1. Map $\hat{T}$ called Arnold Cat.

For example, for $K = 2$, the number of solutions for Eq. (6.2.9) can be derived from Fig. 6.2.1 where $\hat{T}x$ and $\hat{T}^2x$ are plotted versus $x$. The number of solutions equals the number of crossings of the saw-type curve by the square diagonal. For integer $K$, this number is

$$N(m) = K^m = e^{m \ln K} = e^{m h_{KS}} . \tag{6.2.10}$$

Actually, the formula (6.2.10) for the number of periodic orbits can be extended to chaotic systems with good mixing properties. Namely, the number of periodic orbits with period within an interval $(T, T + dT)$ is

$$N(T) \sim \exp(h_{KS}T) .^3 \tag{6.2.11}$$

This result can be compared with the Bowen theorem on the asymptotic equivalence of averaging over the phase space and periodical orbits for a chaotic system. By letting $B(p, q)$ be a physical variable dependent on a point $(p, q)$ in the phase space and $\rho(p, q)$ the distribution function for the chaotic motion in the phase space, it follows that

[3]Rigorous proof of the result (6.2.11) was published in [Mar 69] and [Mar 70].

$$\langle B(p,q)\rangle = \int B(p,q)\rho(p,q)dpdq$$
$$= \lim_{T\to\infty}\frac{1}{\Gamma_A}\sum_{C(T)\in A}\int_{C(T)} B(p_C,y_C)dp_C dq_C\,, \tag{6.2.12}$$

where a small box, $A$, of the volume $\Gamma_A$ in the phase space and all periodic orbits, $C(T)$, of the period $T$ that cross box $A$ are considered. The co-ordinates $p_C, q_C$ belong to a periodic orbit $C$ that crosses box $A$. Equation (6.2.12) is an analog of the ergodic theorem. It means that instead of averaging $B$ over the phase space, the same can be carried out along a fairly long ($T \to \infty$) periodic orbit.[4]

Equations (6.2.11) and (6.2.12) can be used to obtain a qualitative estimate of the probability density $P_R(T)$ for the first recurrence time $T$. For sufficiently large $T$ and small $A$, the set of recurrences in $A$ is similar to the set of periodic orbits that cross $A$. This allows us to consider the distribution of periodic orbits rather than the quasi-periodic ones, which form the set of recurrences in $A$. In other words, it is suggested that not only does an equivalence exists between the averaging over periodic orbits and over stochastic orbits [as stated in (6.2.12)], but a similar equivalence is also present between the distributions of periodic orbits and returns to $A$. Thus, the probability distribution $P_R(T)$ of returns should be proportional to the inverse number of periodic orbits with period $T$ (similar to the micro-canonical Gibbs distribution), that is,

$$P_R(T) \sim 1/N(T) \sim \exp(-h_{KS}T)\,, \tag{6.2.13}$$

in correspondence to (6.2.5).

## 6.3 Non-Ergodicity, Stickiness and Quasi-Traps

It was mentioned above that the Poincaré recurrence theorem does not impose any serious restrictions on the distribution function of cycles $P_R(\tau)$ except for the finiteness of the mean time, $\tau_R$ (6.1.5), and corresponding asymptotics, (6.1.6) and (6.1.7). This restriction is fairly weak since all higher moments

[4]This remarkable result is found in R. Bowen [Bo 72].

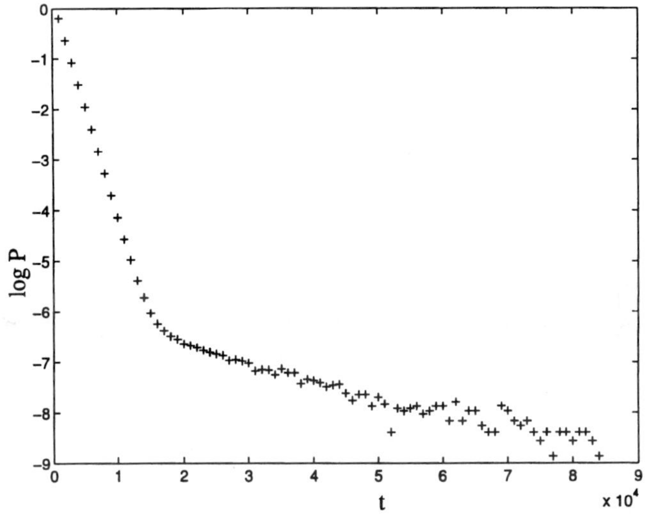

(a)

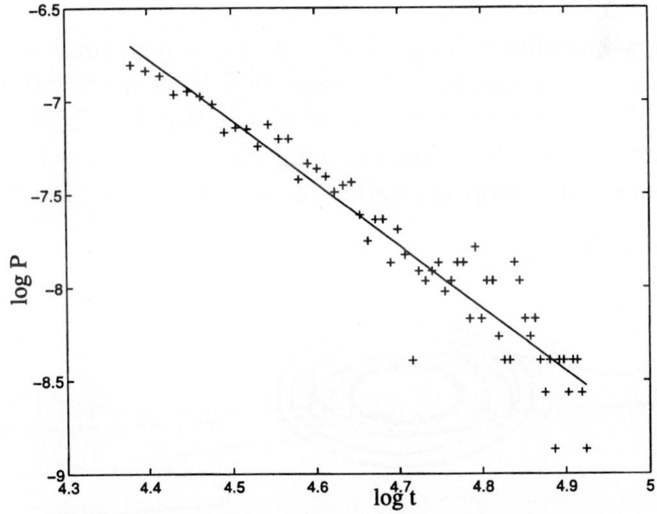

(b)

Fig. 6.3.1. Distribution of Poincaré recurrences for the standard map with $K = 6.908745$: (a) $\log_{10} P$ versus $t$ plot shows crossover from the Poisson law to the power-like distribution; and (b) the power-like tail in log-log plot gives the slope $-3.4$.

$$\langle \tau^n \rangle = \int_0^\infty d\tau \, \tau^n P_R(\tau) \tag{6.3.1}$$

can diverge, beginning with $n_0 > 1$. In this case, one has a situation which is rather unusual for typical thermodynamic-like systems. A discussion of this situation here will be continued in Chapter 7. For the sake of simplicity, one will consider the case when $P_R(\tau)$ has power-like tails in the asymptotics defined by (6.1.6). Many instances of this phenomenon were observed and a demonstration was given in Fig. 6.3.1 for a special case (3.5.16) of the standard map described in Section 3.5. In this case, there exists a hierarchy of islands and sub-islands having the self-similarity property displayed in Table 3.5.2 and characterised by scaling constants (3.5.17). It follows from Fig. 6.3.1 that the characteristic exponent $\gamma$ is the distribution of recurrences $P_R(\tau)$ of (6.1.6) which is equal to

$$\gamma = 3.1 \pm 0.2 \,. \tag{6.3.2}$$

This value means that the second moment of $\tau$ in (6.3.1) can diverge or is approaching divergence.[5]

One now introduces the qualitative notion of a singular, or trapping, zone as an area consisting of the boundary layer around the main island and all hierarchical structures of islands found in the surroundings of the main island. The singular zone can be considered the Hamiltonian analog of a trap or, better to say, quasi-trap (see Fig. 6.3.2)

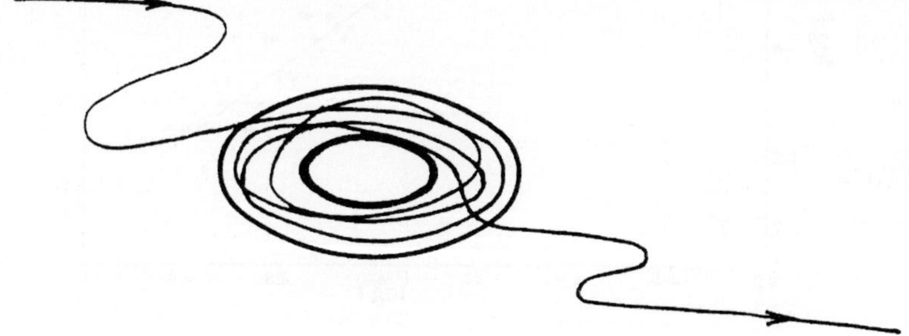

Fig. 6.3.2. A sketch of an orbit that gets stuck (quasi-trap).

[5]The result in Fig. 6.3.1 is taken from [ZEN 97]. Power-like tails in the distribution of recurrence times were previously observed in [CS 84]. See also [C 91] and [ZT 91].

since absolute traps are forbidden in area-preserving dynamics. Considering the returns of a trajectory to the domain $\Delta\Gamma$, one also introduces the escape time from domain $\Delta\Gamma$ which does not exceed the return time.

If

$$P_e(t;\Delta\Gamma) = \int_0^t \psi(\tau;\Delta\Gamma)d\tau \tag{6.3.3}$$

is the probability of exit from domain $\Delta\Gamma$ during time interval $t$ and $\psi(t;\Delta\Gamma)$ is the corresponding probability density for escaping from the $\Delta\Gamma$ in time $t$ within the interval $dt$, the survival probability is then

$$\Psi(t;\Delta\Gamma) = 1 - P_e(t;\Delta\Gamma) = 1 - \int_0^t \psi(\tau;\Delta\Gamma)d\tau = \int_t^\infty \psi(\tau;\Delta\Gamma)d\tau\,, \tag{6.3.4}$$

where the following normalised condition is used:

$$P_e(t \to \infty;\Delta\Gamma) = 1\,. \tag{6.3.5}$$

One can also consider the function $\psi(\tau;\Delta\Gamma)$ as the probability density of particles to be trapped in the domain $\Delta\Gamma$ for a time span of not less than $\tau$. The mean time of trapping is

$$t_s(\Delta\Gamma) = \int_0^\infty \tau\psi(\tau;\Delta\Gamma)d\tau\,. \tag{6.3.6}$$

It is important to stress that all the introduced functions of $\psi$, $\Psi$ and $P_e$ and magnitude $t_s$ are local characteristics of motion, that is, they are defined for an infinitesimal domain $\Delta\Gamma$ and can depend on the shape and location of $\Delta\Gamma$. In other words, $\psi, \Psi, P_e$ and $t_s$ depend only on the properties of the system's trajectories inside $\Delta\Gamma$. The reverse situation is one where the distribution of Poincaré cycles $P(\tau)$ depends on full, infinitely long trajectories and does not depend on $\Delta\Gamma$ for sufficiently small $\Delta\Gamma$. This means that $P(\tau)$ is a global characteristic of the system. Nevertheless, a connection between $P(\tau)$ and $P_e(\tau;\Delta\Gamma)$ can be established for some special cases.

When $\psi(\tau;\Delta\Gamma)$ possesses the asymptotic property

$$\psi(\tau;\Delta\Gamma) \sim t^{-(1+\beta')}, \quad t \to \infty,\ \beta' > 0, \tag{6.3.7}$$

one derives from definition (6.3.4)

$$\Psi(\tau;\Delta\Gamma) \sim t^{-\beta'}, \quad t \to \infty,\ \beta' > 0 \tag{6.3.8}$$

and from (6.3.6),

$$t_s(\Delta\Gamma) \sim \int_{\text{const}}^{\infty} \tau^{-\beta'} d\tau = \begin{cases} \langle\tau\rangle < \infty, & \beta' > 1 \\ \lim\limits_{t\to\infty} \ln t = \infty, & \beta' = 1 \\ \infty & \beta' < 1 \end{cases}. \tag{6.3.9}$$

It follows from (6.3.9) that the mean trapping time is finite only for $\beta' > 1$, while for $\beta \le 1$ the mean trapping time is infinite and the domain $\Delta\Gamma$ works like a real trap in the sense of the time average. It is a trivial condition that

$$t_s < \tau_R < \infty \tag{6.3.10}$$

in the case where the Kac-theorem is valid.

Assuming that the phase space is uniform and has good mixing properties, one can then expect the Poissonian distribution (6.2.5) when the recurrences are $P_R(\tau)$. In the case of non-uniform phase space, there are domains $\Delta\Gamma_s$ of halt in the phase space with islands-around-islands which impose power-like laws for $\psi(\tau;\Delta\Gamma_s)$ and $\Psi(\tau;\Delta\Gamma_s)$. These domains, $\Delta\Gamma_s$, will be called sticky. As can be seen in Table 3.5.2, the closer a trajectory is to a high-order generation island, the longer it sticks to the island, boundary.

One now considers the case of the small domain $\Delta\Gamma$ which is fairly far from the domain $\Delta\Gamma_s$. After departing from $\Delta\Gamma$, a trajectory will occasionally reach the domain $\Delta\Gamma_s$ and then return to $\Delta\Gamma$ repeatedly. For fairly short periods of time, $P(\tau)$ resembles the Poissonian distribution and for fairly long periods $\tau \gg \tau_0$, the stickiness of the domain $\Delta\Gamma_s$ imposes the anomalous asymptotics of the kind found in (6.3.7) and (6.3.8). It follows that

$$P_R(\tau) \sim \tau^{-(1+\beta')}, \quad \tau \gg \tau_0 \tag{6.3.11}$$

where $\tau_0$ is the estimated time required for a particle to reach $\Delta\Gamma_s$ if it starts at $\Delta\Gamma$. In (6.3.8), the exponent $\beta'$ depends on the choice of $\Delta\Gamma_s$. The exponent $\gamma$ in (6.1.6) does not depend on the choice of $\Delta\Gamma$ for the Poincaré recurrences. Thus, the existence of a singular sticky domain imposes its own dominant asymptotics that reveal themselves in the expression (6.3.11) for the anomalous, non-Poissonian distribution. These comments lead to the following conjecture: when the only singular zone with the characteristic exponent $\beta'$ [see (6.3.8)] for the survival probability exists, the asymptotic relation

$$\psi(t, \Delta\Gamma) \sim P_R(t) \sim t^{-\gamma}, \quad t \to \infty \tag{6.3.12}$$

with

$$\gamma = 1 + \beta' \tag{6.3.13}$$

and $\Delta\Gamma$ is taken in the singular zone. In the case of several singular zones, one can expect different intermediate asymptotics such that the larger the time period, the smaller the $\gamma$. These situations urge one to introduce fractal, or even multi-fractal, descriptions of properties such as the time of escapes and returns, which are considered in this chapter.

Special consideration should be given to the ergodicity property in the presence of singular zones $\Delta\Gamma_s$. Consider the entire phase volume of the system extracting the part which belongs to the islands. In this remaining part, the ergodic motion with finite measure of chaotic orbits is assumed.

The fact that the distribution functions (6.3.12) have self-similarity properties has led to an unusually slow convergence in the condition associated with the ergodic property or the existence of the limit

$$\bar{f}(x(t)) \equiv \lim_{t\to\infty} \frac{1}{t} \int_0^t f(x(\tau))d\tau = \langle f(x(t))\rangle = \langle f(x(0))\rangle = f_0 \,, \tag{6.3.14}$$

where $f$ is an integrable function of the dynamical variables $x$.

The usual way of observing the property (6.3.14) is to find the corresponding values $\bar{f}$ and $\langle f\rangle$ and to compare them. For example, phase averaging $\langle f\rangle$ can be performed by observing $N$ orbits at time span $t_N$ each. One can expect that if

$$Nt_N = t \tag{6.3.15}$$

and $t_N$ and $N$ are sufficiently large, the results for $\bar{f}$ and $\langle f \rangle$ in (6.3.14) will be approximately the same. The self-similarity of the random process of chaotic dynamics and the existence of quasi-traps in the phase space of a system can give rise to new problems. The most typical questions asked are: what is the time scale $t_0$ and what is the number of orbits $N_0$ required such that for $t_N > t_0$ and $N > N_0$ the value of $\bar{f}$ is close to $\langle f \rangle$. If the pair $(t_0, N_0)$ exists, then one can use condition (6.3.15) in various ways. It is only for systems with good mixing properties that one can prove the fast convergence of $\bar{f}$ to $\langle f \rangle$ and the existence of the characteristic values of $(t_0, N_0)$. In fact, this is generally not the case in Hamiltonian systems and slow convergence in (6.3.14) for dynamical chaos can create serious difficulties for chaos theory.[6]

## 6.4 Renormalisation Formulas for the Exit Time Distribution

It was mentioned above that the exit time probability distribution $P_e(t; \Delta\Gamma)$ depends on the location of the domain $\Delta\Gamma$. If, for example, $\Delta\Gamma$ is taken in the domain $\Delta\Gamma_s$ of a singular zone around an island, then due to (6.3.3) and (6.3.12), one obtains

$$\dot{P}_e(t; \Delta\Gamma_s) = \psi(t; \Delta\Gamma_s) \sim P_R(t)\,, \quad (t \to \infty) \tag{6.4.1}$$

where the last relation is valid only asymptotically and does not depend on $\Delta\Gamma_s$. Equation (6.3.12) has two advantages:

1. It presents the possibility of finding the asymptotics for the distribution function of recurrences $P_R(t)$ using the local information of singular zone properties.
2. Due to the space-time similarity of domain $\Delta\Gamma_s$ in the phase space (as discussed in detail in Section 3.5), it enables the application of renormalisation methods to $\psi(t; \Delta\Gamma_s)$.

Assuming that a singular zone, $\Delta\Gamma_s$, is an annulus surrounding the island (as shown in Fig. 6.4.1) and $q$ islands of the first generation are

[6] For more discussion on this topic, see [ASZ 91] and [ZEN 97].

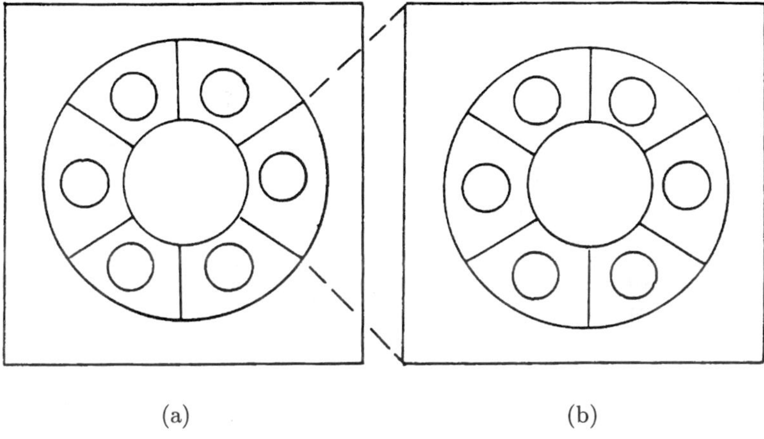


Fig. 6.4.1. A model of the singular zone near the island boundary: (a) boundary layer as an annulus and its partition; and (b) self-similar rescaling.

present in the annulus ($q = 6$ in Fig. 6.4.1), one then divides the annulus into $q$ equal parts with each containing a sub-island and performs the same type of partitioning on each part (this is the transition from $a$ to $b$ in Fig. 6.4.1). This process can be continued. The result resembles Fig. 5.1.1. Recall that the self-similarity properties for the islands in the vicinity of the boundary layers (annuluses) are

$$\delta S_{k+1} = \lambda_S \delta S_k \,, \tag{6.4.2}$$

and for periods of rotation around the islands,

$$T_{k+1} = \lambda_T T_k \,, \tag{6.4.3}$$

as mentioned in Section 3.5 [see (3.5.11)].

After transferring the properties of self-similarity, (6.4.2) and (6.4.3), to the exit time probability density $\psi(t, \Delta\Gamma_s)$, one turns to a consideration of the different domains, $\Delta\Gamma_k$, and the partition shown in Fig. 6.4.1 which correspond to a generation of order $k$. This means that

$$\Delta\Gamma_{k+1} \sim \lambda_s \Delta\Gamma_k \,. \tag{6.4.4}$$

Thus, the exit time distribution, $P_e(t; \Delta\Gamma_k)$, and the corresponding survival distribution, $\Psi(t; \Delta\Gamma_k)$, satisfy the scaling equation which follows from a renormalisation of time:

$$\Psi(t - t_{k+1}; \Delta\Gamma_{k+1}) = 1 - P_e(t - t_k; \Delta\Gamma_{k+1}) = \Psi(t - t_k; \Delta\Gamma_k)$$
$$= [1 - P_e(t - t_k; \Delta\Gamma_k)] \qquad (6.4.5)$$

where each of the functions, $P_e(t - t_j; \Delta\Gamma_j)$, is normalised for each corresponding domain, $\Delta\Gamma_j$ and $t_k \sim \lambda_T^k$. A similar equation can also be written for the density distribution:

$$\psi(t - t_{k+1}; \Delta\Gamma_{k+1}) = \psi(t - t_k; \Delta\Gamma_k) . \qquad (6.4.6)$$

Equations (6.4.5) and (6.4.6) are valid only for the tail part of the distribution functions. Using (6.3.7), we obtain the following shift formula in the logarithmic scale:

$$\ln[\psi(\ln(t - t_k) - \ln \lambda_T; \Delta\Gamma_{k+1})] = \ln[\psi(\ln(t - t_k); \Delta\Gamma_k)] . \qquad (6.4.7)$$

The formula (6.4.7) should be interpreted in the following way: the domain $\Delta\Gamma_k$ is in the annulus of the $k$-th generation in Fig. 6.4.1 and the asymptotic behaviour of the exit time density distribution function is $\psi(t; \Delta\Gamma_k)$, or more precisely, $\ln \psi$ versus $\ln t$. The same type of dependence exists in domain $\Delta\Gamma_{k+1}$ with substracted $\Delta\Gamma_k$, and the corresponding power-like dependences should be the same up to a time shift by $\ln \lambda_T$. Such a shift can be observed in simulations (see Fig. 6.4.2) where the value of the shift is of the order $\ln \lambda_T = \ln 8 \approx 2.2$.

One now presents a demonstration on how to obtain the exponent $\beta'$ for the distributions (6.3.7) and (6.3.8) by using Eq. (6.4.5). For the sake of simplicity, the interval

$$\delta t_k = ((1 + \delta^-)\lambda_T^{k-1}, (1 + \delta^+)\lambda_T^k) \qquad (6.4.8)$$

is introduced where $|\delta^\pm| < \lambda_T$ are some numbers. One also considers

$$t - t_k \in \delta t_k \qquad (6.4.9)$$

together with $t_k \sim \lambda_T^{k-1}$ in order to specify the area of consideration for functions $\Psi(t - t_k; \Delta\Gamma_k)$. From the definitions of $\Psi(t - t_k; \Delta\Gamma_k)$ and $\Psi(t - t_{k+1}; \Delta\Gamma_{k+1})$, it follows that

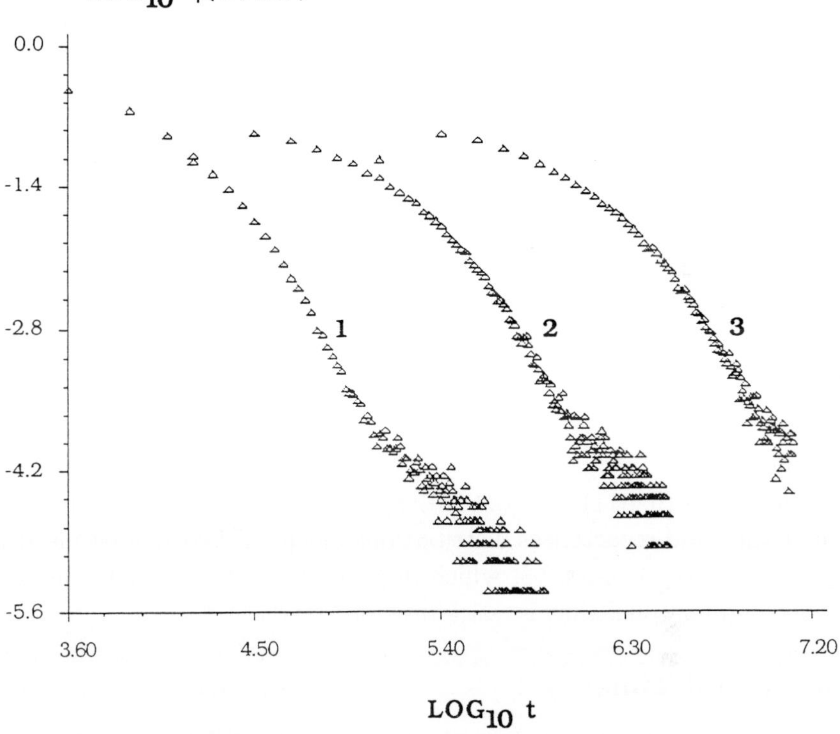


Fig. 6.4.2. Exit time distributions for domains taken in the boundary layers for the first, second and third generations (curves 1, 2 and 3, respectively).

$$\begin{aligned}\Psi(t - t_{k+1}; \Delta\Gamma_{k+1}) &= \frac{\text{const.}}{(t - t_{k+1})^{\beta'}} \\ &= \Psi\left(\lambda_T\left(\frac{t}{\lambda_T} - t_k\right); \Delta\Gamma_k \cdot \frac{\Delta\Gamma_{k+1}}{\Delta\Gamma_k}\right) \\ &= \Psi\left(\lambda_T\left(\frac{t}{\lambda_T} - t_k\right); \lambda_S \Delta\Gamma_k\right) \\ &= \frac{\lambda_T}{\lambda_S}\, \frac{1}{\lambda_T^{\beta'}} \Psi(t - t_k; \Delta\Gamma_k)\,, \end{aligned} \tag{6.4.10}$$

where the argument in $\Psi$ in the last expression satisfies $t - t_k \in \delta t_k$. Applying (6.4.5) to (6.4.10), one derives the equation $\lambda_S \lambda_T^{\beta'-1} = 1$ or

$$\beta' = 1 + |\ln \lambda_S| / \ln \lambda_T = 1 + \mu \,, \tag{6.4.11}$$

where the important exponent $\mu$ has been introduced:

$$\mu = |\ln \lambda_S| / \ln \lambda_T \,. \tag{6.4.12}$$

$\mu$ is called a transport exponent for reasons that will soon become clear. From the formulas (6.4.11) and (6.3.13), one also obtains an important relation,

$$\gamma = 2 + \mu \,, \tag{6.4.13}$$

which defines the recurrences distribution exponent $\gamma$ [see (6.3.12)].

The results (6.4.11) and (6.4.13) express the characteristic exponents of exit time and recurrences distributions using the two space-time scaling parameters, $\lambda_S$ and $\lambda_T$, which describe the intrinsic self-similarity of the singular zone, that is, stickiness of islands.

## 6.5 Fractal Time

One is now ready to introduce the notion of fractal time in the same manner as what M. Shlesinger and his co-authors have done.[7] A trivial and quick way is to consider a set of events ordered in time and to apply a notion of fractal dimension to the set of time instants, say $\{t_j\}$. However, this method is too formal and not representative enough. In the previous section, it was shown that for chaotic trajectories the distribution of exit time and Poincaré recurrences possesses the self-similarity property in large time asymptotics, and that corresponding power-like tails occur in (6.3.12). It also was mentioned that a power-like tail in the time events distribution leads to the divergence in moments.

The idea of considering the random processes of infinite moments is historically related to the so-called St. Petersburg Paradox of Nicolas

[7]For a detailed description and applications to different statistical problems, including the material problem, see [MS 84], [Sh 88] and [SSB 91]. The application to dynamical chaos can be found in [Za 94–2] and [Za 95].

Bernoulli. It is based on a special scaling known as the "Bernoulli scaling".[8] When the events $\{\tau_j\}$ are scaled as

$$\{\tau_j\} = \{\tau, b\tau, b^2\tau, \ldots\}\,, \tag{6.5.1}$$

the probability of having an event $\tau_j = b\tau$ is scaled as

$$P_j(\tau) = c^j p(b^j \tau)\,, \tag{6.5.2}$$

with some scale parameters $b, c$. Expressions (6.5.1) and (6.5.2) can be interpreted as follows: there are different ways to escape from the domain $\Delta\Gamma$ and each $j$-th way is described by the probability density $P_j(\tau)$. Hence the probability density of escaping in time $t \geq \tau$ is

$$\psi(\tau) = \sum_{j=0}^{\infty} p_j(\tau) = \sum_{j=0}^{\infty} c^j p(b^j \tau) \tag{6.5.3}$$

with the normalisation condition being

$$\int_0^{\infty} \psi(\tau) d\tau = 1\,. \tag{6.5.4}$$

In the St. Petersburg Paradox, Bernoulli had considered the tossing of a coin. The $j$-th event is a set of $j$ unsuccessful tail flips until the first head is tossed and the corresponding probability is $p_j = (1/2)^{j+1}$. The winning award was $d^j$ so that the mean winning is

$$\langle d \rangle = \sum_{j=0}^{\infty} d^j p_j = \frac{1}{2} \sum_{j=0}^{\infty} (d/2)^j\,, \tag{6.5.5}$$

which diverges if $d \geq 2$. Similarly, it is possible that $\langle d \rangle < \infty$ but $\langle d^2 \rangle = \infty$ if $d \geq 2^{1/2}$.

Fractal properties of the exit time distribution can be demonstrated using a model for (6.5.3) in the form of

$$\psi(t) = \frac{1-a}{a} \sum_{j=0}^{\infty} (ab)^j \exp(-b^j t) \tag{6.5.6}$$

---

[8] For more information on the St. Petersburg Paradox and Bernoulli scaling, see [MS 84].

which incorporates the Bernoulli scaling. The expression (6.5.6) represents a superposition of Poissonian distributions with different time scales. It is easy to derive from (6.5.3) that

$$\psi(t) = c\psi(bt) + p(t)\,, \tag{6.5.7}$$

which can be considered as an equation to determine the behaviour of $\psi(t)$ as a function of $t$. The same type of equation occurs in the renormalisation group theory of phase transitions. It shows that the notion of fractal time can be considered by applying the general scheme of genera-lised fractal variable and the corresponding renormalisation equation.

It is possible to find that asymptotically

$$\psi(t) \sim 1/t^{1+\beta''}\,, \quad (t \to \infty) \tag{6.5.8}$$

with

$$\beta'' = \ln a / \ln b \tag{6.5.9}$$

where $a$ and $b$ are the same as in (6.5.6) (see Appendix 2). The result is similar to (6.4.11), but it is too formal and its relation to dynamical models is not yet established. In the next section, a more general approach will be introduced.

## 6.6 Fractal and Multi-Fractal Recurrences

For Hamiltonian systems with chaotic dynamics, the motion is not ergodic in the full phase space and one therefore needs to subtract a (multi-) fractal set of islands to obtain a domain with ergodic trajectories. The islands form a singular part of the phase space. The behaviour of the trajectories near an island boundary layer was discussed in Chapter 3. It was shown that for some values of the control parameter, the set of islands is a fractal object which imposes a powerful distribution of different time-dependent characteristics in large time asymptotics. More specifically, the boundary of the island is sticky, with that of the sub-island being more sticky and so forth. Consequently, the fractal space-time properties of the trajectories should be examined. One encounters a new situation in which: (i) fractal properties exist simultaneously in space and time; and (ii) the multiplicity of the resonance sets that generate islands corresponds to a multi-fractal construction of the trajectories rather than to a fractal one.

This section describes the multi-fractal time and the corresponding spectral function of dimensions. In general, for chaotic dynamics, the fractal time cannot be introduced without a consideration of the space structure. For that reason, space-time coupling is non-trivial and the same is true of the spectral function of the fractal dimensions. One needs to extend the methods described in Section 5.4 in such a way that they permit a consideration of the phase space and time instants of the events related to a chaotic trajectory as a (multi-) fractal object. As an event, one can again take the particle trajectory that enters a cell, which is a part of the annulus in Fig. 6.4.1, and spends some time there before escaping from the cell. This process was described in Section 6.4. Also, for some values of the control parameter $K$ in the standard map and the web-map, the self-similarities of the islands' area and the periods of their last invariant curves were demonstrated in Tables 3.5.1 and 3.5.2. However, a more general way of considering the hierarchical structures is needed.

Using space-time partitioning, which was introduced in Fig. 6.4.1 and resembles the Serpinsky carpet (Fig. 5.1.1), the central square is assumed to be an island of zero-generation. The island is then surrounded by an annulus which represents the boundary island layer. It consists of $g_1$ ($g_1 = 8$ in Fig. 5.1.1) sub-islands of the first generation (dashed smaller islands in Fig. 5.1.1). One can partition the annulus by $g_1$ domains in the same manner as in Fig. 6.4.1 so that each of them includes exactly one island of the first generation. Each of the first-generation island is then surrounded by an annulus of the second generation. The process is repeated, after which the segments are squared before arriving at Fig. 5.1.1. On the $n$-th step, the structure is described by a "word":

$$w_n = w(g_1, g_2, \dots, g_n)\,. \tag{6.6.1}$$

The full number of islands on the $n$-th step is

$$N_n = g_1 \cdots g_n \tag{6.6.2}$$

and any island from the $n$-th generation can be labelled as

$$u_i^{(n)} = u(i_1, i_2, \ldots, i_n) \quad 1 \leq i_j \leq g_j, \ \forall\, j\,. \tag{6.6.3}$$

The time that a particle spends in the boundary layer of an island is then introduced. This time,

$$T_i^{(n)} = T(u_i^{(n)})\,, \tag{6.6.4}$$

carries all the information on the $n$-th generation islands (6.6.1) to (6.6.3). By introducing a residence time for each island boundary layer, a new situation, compared to the plain Sierpinsky carpet or a plain fractal situation, arises because of the non-triviality of space-time coupling. In fact, an additional parameter responsible for the temporal behaviour is attached to a simple geometric construction similar to the Cantor-set.

A simplified situation corresponds to the exact self-similarity of the construction described above, that is,

$$\begin{aligned} S_i^{(n)} &= S^{(n)} = \lambda_S^n \cdot S^{(0)}\,, \quad (\forall\, i) \\ T_i^{(n)} &= T^{(n)} = \lambda_T^n \cdot T^{(0)}\,, \quad (\forall\, i) \end{aligned}\,, \tag{6.6.5}$$

where $S_i^{(n)}$ is the area of an island $u_i^{(n)}$ and $T_i^{(n)}$ is introduced in (6.6.4). The expressions (6.6.5) correspond to equal areas and residence times for all islands of the same generation. Two scaling parameters, $\lambda_S$ and $\lambda_T$, represent the existence of the exact self-similarity in space and time, respectively. Such a situation was described in Section 3.5 for the standard map and the web-map with

$$\lambda_S < 1\,, \quad \lambda_T > 1\,. \tag{6.6.6}$$

In addition to (6.6.6), there is a self-similarity in the islands' proliferation, that is,

$$g_n = \lambda_g^n g_0\,, \quad \lambda_g \geq 3\,. \tag{6.6.7}$$

It follows from (6.6.2) and (6.6.7) that

$$N_n = \lambda_g^n g_0 = \lambda_g^n \tag{6.6.8}$$

if one begins with a single island ($g_0 = 1$). It is useful to introduce a "residence frequency",

$$\omega_i^{(n)} = 1/T_i^{(n)}, \tag{6.6.9}$$

with the self-similarity property

$$\omega_i^{(n)} = \omega^{(n)} = \lambda_T^{-n}\omega^{(0)}. \tag{6.6.10}$$

Turning now to a consideration of the partition which corresponds to conditions (6.6.1) to (6.6.4) with a simplification of (6.6.7), the $n$-th level of the partition corresponds to that of the islands' hierarchy. This means that each space bin has an area $S_i^{(n)}$ and a co-joint residence time $T_i^{(n)}$ [both are defined in (6.6.5)] whose values do not depend on $i$. The elementary probability of spending time $T_i^{(n)}$ in the domain $S_i^{(n)}$ can be presented in a simple form as

$$P_i^{(n)} \equiv P_{i_1,i_2,\ldots,i_n} = C_n\omega_i^{(n)}S_i^{(n)}, \quad \forall\, 1 \le i_j \le g \tag{6.6.11}$$

where $C_n$ is a normalisation constant. Calling $P_i^{(n)}$ an elementary bin-probability and using (6.6.5), (6.6.6) and (6.6.9), one can rewrite (6.6.11) as

$$P_i^{(n)} = C_n(\lambda_S/\lambda_T)^n, \quad (\forall\, i). \tag{6.6.12}$$

Having derived (6.6.12), one can consider different sums and partition functions. For example, in the sum

$$\begin{aligned} Z^{(n)} &= \sum_{i_1,\ldots,i_n} P_i^{(n)} = C_n \sum_{i_1,\ldots,i_n} \exp[-n(|\ln\lambda_S| + \ln\lambda_T)] \\ &= \sum_{i_1,\ldots,i_n} \exp[-n(|\ln\lambda_S| + \ln\lambda_T) + n\psi_n] = 1, \end{aligned} \tag{6.6.13}$$

where the "free energy" density

$$\psi_n = \frac{1}{n}\ln C_n \tag{6.6.14}$$

is introduced, the number of terms in (6.6.13) follows from (6.6.8). Therefore, in the limit $n \to \infty$, one obtains

$$\lim_{n\to\infty}\psi_n = \psi = |\ln\lambda_S| + \ln\lambda_T - \ln\lambda_g. \tag{6.6.15}$$

A more precise formulation of the result in (6.6.15) is that the sum in (6.6.13) diverges if $\psi_n > \psi$, converges to zero if $\psi_n < \psi$, and converges to one if $\psi_n = \psi$.

The goal of these manipulations is to describe the method of partitioning and how to operate using a corresponding sum, such as $Z^{(n)}$, in the case where a bin of the partition has two features related to the phase space location and volume, and to the residence time. On the basis of this information, one can introduce a multi-fractal spectrum for the recurrence time set.

## 6.7 Multi-Fractal Space-Time and Its Dimension Spectrum

It was mentioned in Section 3.5 that chaotic dynamical systems with rich sets of islands have a multi-fractal rather than fractal space-time structure. This section introduces a spectral function of dimensions which is analogous to Section 5.4.

Following a method common in statistical mechanics, a partition function in the form of

$$Z_{\text{discr}}^{(n)}\{\lambda_T, \lambda_S; q\} = \sum_{i_1, i_2, \ldots, i_n} \left(\omega_i^{(n)} S_i^{(n)}\right)^{\gamma q} \qquad (6.7.1)$$

is introduced.

Here one uses the space-time partitioning probability $\omega_i^{(n)} S_i^{(n)}$ derived in (6.6.11). Though similar to (6.6.13), it has a more general character. Considering a multi-scaling situation, it is assumed that the real elementary probability of occupying a bin has the same scaling dependence as (6.6.11) of up to a power of $\gamma$, and that there are different values of $\gamma$ in the sum. With the exponent $q$, one can consider the different moments of the elementary bin probability. In particular, for $q = 0$, one can simply obtain the number of bins. Replacing summation by integration and having

$$Z^{(n)}\{\lambda_T, \lambda_S; q\} = \int d\gamma \rho(\gamma) \left[\omega^{(n)} S^{(n)}\right]^{-f(\gamma)+\gamma q}, \qquad (6.7.2)$$

the density of the space-time bins is introduced:

$$dN^{(n)}(\gamma) = d\gamma\, \rho(\gamma) \left[\omega^{(n)} S^{(n)}\right]^{-f(\gamma)} . \tag{6.7.3}$$

The function $f(\gamma)$ is a spectral function of the space-time dimension-like characteristics or, simply, dimensions. The distribution density $\rho(\gamma)$ is a slow function of $\gamma$. To be more accurate, it is also assumed that the bin-probability $\omega^{(n)} \cdot S^{(n)}$ depends on $\gamma$. This is because for different island-sets, the bins have different structures. Nevertheless, the dependence of $\omega^{(n)} \cdot S^{(n)}$ on $\gamma$ is dependence when compared to the exponential dependence in (6.7.3).

When (6.6.5) and (6.6.10) are used, (6.7.2) is transformed into

$$Z^{(n)}\{\lambda_T, \lambda_S; q\} = \int d\gamma \rho(\gamma) \exp\{-n[\gamma q - f(\gamma)](|\ln \lambda_S| + \ln \lambda_T)\}, \tag{6.7.4}$$

where $\lambda_S$ and $\lambda_T$ are slow functions of $\gamma$. For $n \to \infty$, the standard steepest descent procedure yields

$$Z^{(n)}\{\lambda_T, \lambda_S; q\} \sim \exp\{-n[\gamma_0 q - f(\gamma_0)] \cdot (|\ln \lambda_S| + \ln \lambda_T)\}, \tag{6.7.5}$$

where $\gamma_0 = \gamma_0(q, \lambda_S, \lambda_T)$ and it satisfies the equation

$$q = f'(\gamma_0) . \tag{6.7.6}$$

On the other hand, recall that for $q = 0$, expression (6.7.1) defines $Z^{(n)}_{\text{discr}}\{\lambda_T, \lambda_S; 0\}$ as the number of bins. For the one-scale situation, this number can be derived from (6.6.8) as $\lambda_g^n$. In the case of the multifractal, a power of $\lambda_g^n$ can be written by introducing a generalised dimension, $D_q$, which was carried out in Section 5.4:

$$Z^{(n)}\{\lambda_S, \lambda_T; q\} \sim \lambda_g^{-n(q-1)D_q} \sim \exp\{-n \ln \lambda_g \cdot (q-1) D_q\} . \tag{6.7.7}$$

The scaling parameter $\lambda_g$ defines a coefficient of the proliferation of space-time bins and one can therefore consider

$$\lambda_g = \lambda_g(\lambda_T, \lambda_S) , \tag{6.7.8}$$

that is, dynamical systems with only two independent scaling parameters will be examined. This restriction is not important and, if necessary, it can be very easily lifted.

A comparison of (6.7.7) and (6.7.5) yields

$$(q-1)D_q \cdot \ln \lambda_g = [\gamma_0 q - f(\gamma_0)](\ln \lambda_T + |\ln \lambda_S|)\,. \tag{6.7.9}$$

In some limit cases, $\lambda_g$ in (6.7.8) should satisfy the fullowing conditions:

$$\lambda_g = \begin{cases} |\lambda_S|\,, & \text{if } \lambda_T = 1 \\ \lambda_T\,, & \text{if } \lambda_S = 1\,. \end{cases} \tag{6.7.10}$$

One then arrives at the standard situation of one-parametric scaling. For a general situation $\lambda_g$, $\lambda_T, \lambda_S \neq 1$, (6.7.9) is rewritten in its final form as

$$(q-1)D_q = \frac{\ln \lambda_T}{\ln \lambda_g}(1+\mu)[\gamma_0 q - f(\gamma_0)]\,, \tag{6.7.11}$$

where the parameter

$$\mu = |\ln \lambda_S| / \ln \lambda_T \tag{6.7.12}$$

was introduced in (6.4.12) and termed a transport exponent.

It follows from (6.7.11) that for $q = 0$,

$$D_0 = \frac{\ln \lambda_T}{\ln \lambda_g}(1+\mu) \cdot f(\gamma_0)\,. \tag{6.7.13}$$

This means that there is now no simple connection between the dimension $D_0$ and the spectral function. The regular formula

$$D_0 = f(\gamma_0) \tag{6.7.14}$$

appears only in the case of (6.7.10) when a multi-fractal structure exists only in space or in time. For $q = 1$, the use of (6.7.6) and (6.7.11) yields

$$D_1 = \gamma_0(1) \cdot \frac{\ln \lambda_T}{\ln \lambda_g}(1+\mu)\,, \tag{6.7.15}$$

where the value $\gamma_0(1) = \gamma_0(q=1)$ can be obtained from Eq. (6.7.6):

$$f'(\gamma_0(q=1)) = 1\,.$$

In (6.7.9), the generalised dimension $D_q$ was expressed through the spectral function $f(\gamma)$ (which was the case in Section 5.4). Nevertheless, the formulas (6.7.9), (6.7.11), (6.7.13) and (6.7.15) have shown that a knowledge of the spectral function is not sufficient to describe a typical dynamical system, and additional information on the system's structure in space and time is therefore necessary. A simplification can be made in cases where $\lambda_g, \lambda_S$ and $\lambda_T$ for some special values of the control parameter are known.

## 6.8 Critical Exponent for the Poincaré Recurrences

This section examines the partitioning introduced in Section 6.6 (see Fig. 5.1.1) and the recurrences or escapes from the boundary island layer of the $n$-th generation. When normalised to the unit of the probability (6.6.11) for a bin to be occupied by a particle, the corresponding "number of states" can be written as

$$Z_r^{(n)} = \sum_{i_1,\dots,i_n} \frac{1}{S_i^{(n)}\omega_i^{(n)}} = \sum_{i_1,\dots,i_n} (\lambda_T/\lambda_S)^n \,. \tag{6.8.1}$$

Instead of (6.8.1), one considers a more general expression:

$$Z_r^{(n)}(q) = \sum_{i_1,\dots,i_n} \frac{1}{S_i^{(n)}\left[\omega_i^{(n)}\right]^q} = \sum_{i_1,\dots,i_n} \lambda_T^{nq}/\lambda_S^{-n} \,. \tag{6.8.2}$$

Using (6.6.8), the following estimation is derived:

$$Z_r^{(n)}(q) \sim \exp\{n(q \ln \lambda_T + |\ln \lambda_S| + \ln \lambda_g)\} \,. \tag{6.8.3}$$

The expression can be simplified if $\lambda_g = \lambda_T$ and the proliferation coefficient for the number of islands coincides with that of the increase in the circulating period around the islands. Hence

$$Z_r^{(n)}(q) \sim \exp\{n[(q+1)\ln \lambda_T + |\ln \lambda_S|]\} \,. \tag{6.8.4}$$

This expression is finite if

$$q \le q_c = -(|\ln \lambda_S| + \ln \lambda_T)/\ln \lambda_T = -(1+\mu) \,. \tag{6.8.5}$$

The result thus obtained has a remarkable interpretation.

From the definition (6.8.2), $Z_r^{(n)}(q)$ can be considered time-moments of order $q$ for the escape or recurrence sum of the states $Z_r^{(n)}$. It is only finite if $|q| \le |q_c|$. This means that the probability density for recurrences [see (6.1.6)] should possess the asymptotics

$$P_R(t) \sim t^{-1+q_c} = t^{-1-(1+\mu)} = t^{-2-\mu} = t^{-\gamma} \tag{6.8.6}$$

in order to have finite moments of order $q > 0$. The result in (6.8.6) gives rise to

$$\gamma = 2 + \mu \tag{6.8.7}$$

and coincides with the expression (6.4.13) in Section 6.4 from a different consideration.

## Conclusions

1. The distribution of Poincaré recurrences $P_R(\tau)$ is a way of characterising chaotic trajectories. For typical chaotic systems, $P_R(\tau)$ is of a Poissonian type:

   $$P_R(\tau) = \frac{1}{\langle\tau\rangle} \exp\left(-\frac{\tau}{\langle\tau\rangle}\right).$$

   It has a mean recurrence time $\langle\tau\rangle$ reciprocal to Kolmogorov-Sinai entropy.

2. M. Kac proved that for the Hamiltonian, finite dynamics with non-zero measure $\langle\tau\rangle < \infty$. Consequently, for the power-like asymptotics,

   $$P_R(\tau) \sim \tau^{-\gamma}\,, \quad (\tau \to \infty)$$

   should be

   $$\gamma > 2\,.$$

3. The case of the non-exponential behaviour of $P_R(\tau)$ is related to the anomalous transport with non-Gaussian time dependence of moments. In some special cases, a connection can be found:

   $$\gamma = 2 + \mu$$

   where $\mu$ is the transport exponent for the second moment.

4. It is possible to find an expression for $\mu$ for certain situations which have a scaling property:

$$\mu = |\ln \lambda_S| / \ln \lambda_T \, .$$

$\lambda_S, \lambda_T$ are the scaling parameters for the area $S$ and period $T$ of the last invariant curve of the islands hierarchy in the singular zone (boundary layer).

5. The singular zone can be considered a quasi-trap in the phase space and the fractal time concept can be applied to the quasi-trap.

6. The multi-fractal spectral function can be introduced for the Poincaré recurrences set. The corresponding partition should be carried out in space and time.

*Chapter 7*

# CHAOS AND FOUNDATION OF STATISTICAL PHYSICS

## 7.1 The Dynamical Foundation of Statistical Physics

An informal indication of the problem in explaining the meaning of the foundation of statistical physics can be seen in the presence of two types of processes: (i) time-irreversible macroscopic processes which obey the thermodynamic, or kinetic, laws; and (ii) time-reversible microscopic processes which obey, say, the Newton and Maxwell equations of motion. The great difference between both descriptions of the processes in nature is not clearly understood and an acceptable explanation on the origin of irreversibility is still lacking. The formal definition of the problem is to derive a kinetic equation which describes an irreversible evolution which begins with a reversible Hamiltonian equation. This approach is considered as a derivation of the statistical and corresponding thermodynamic laws from the first principles.

Boltzmann's kinetic theory is considered the first successful attempt to unite, in a formal way, the statistical and dynamical description of a system. Boltzmann's contemporaries failed to appreciate the fact that his idea was essentially based on the existence of two different scales in time and space: the smaller $\tau_c$, which is applicable to a purely dynamical process of collisions, and the larger $\tau_d \gg \tau_c$ applicable to the statistical process of a system's evolution described by a kinetic equation. The resulting equilibrium distribution, if it exists, would thus correspond to the thermodynamical state. However, non-equilibrium states are also possible.

Boltzmann's ideas and notions of the kinetic equation and H-theorem were not well-received by his contemporaries and were heavily criticised, in particular, by Zermelo. It was only after his death, when a thorough statistical analysis of his theory had been made by P. Ehrenfest and T. Ehrenfest and the first numerical analysis was performed (the "urn model" by Ehrenfest), that the validity of Boltzmann's ideas was acknowledged. The critical comments by Zermelo, known as the Zermelo's paradox, were based on the Poincaré Recurrence Theorem and contained many flawed statements (see Section 6.1 and footnotes [1] and [2] in Chapter 6).

Subsequently, many scientists were concerned with the problem of statistical physics foundation, defined above as the problem of obtaining the system's kinetic description based on the first principles, that is, the system's Hamiltonian, and possibly on some accurately formulated supplementary conditions. This has been successfully achieved in different ways and with varying degrees of generalisation. It was shown that the random phase approximation and the Gaussian nature of collision micro-processes or other equivalent conditions played the role of statistical element. This had to be added to the Hamiltonian dynamical equations in order to obtain a kinetic equation. The averaging over micro-random variables resulted in a reduced description of the system in the space of generalised actions or moments.[1]

Despite the formal success of the derivation of the kinetic equation from a Hamiltonian, this approach is unsatisfactory since a very strong assumption, the randomness of some variables, that is, to put "by hands", needs to be introduced. The theory of dynamical chaos has altered the perception of the possibility of statistical laws foundation since the dynamical trajectories, being solutions of the deterministic equations of motion, may resemble the curves which represent a random process. When he was analysing the foundations of quantum mechanics,

[1]The literature on kinetic equations derived in the random phase approximation is fairly extensive. The books by I. Prigogine [Pr 62] and R. Balescu [Ba 75] are highly recommended. Many original discussions can be found in the book by M. Kac [Ka 58]. The role of chaos in the derivation of the kinetic equation is discussed in great detail in [Za 85].

Einstein made a well-known remark: "God doesn't play dice with the world" (see [Kl 70]). These words can be further amplified from the standpoint of classical physics: God does not have to play dice since dynamical equations can in themselves give rise to stochastic processes if certain simple restrictions for the parameters and initial conditions are applied. The use of the phenomenon of dynamical chaos in statistical physics foundation began with Krylov and has since been used in classical and quantum physics.[2] Strictly speaking, the presence of dynamical chaos in a system is not sufficient for the foundation of statistical physics laws, or more accurately, this foundation has not been built so far. The major obstacle in substantiating statistical laws lies in the way dynamical chaos is structured in real Hamiltonian systems and how the real kinetics look like in such systems. It was shown in Chapter 6 that the distribution of the Poincaré recurrences is time- and space-fractal. In the phase space, one encounters domains such as quasi-traps. This chapter discusses the difficulties of using the properties of dynamical chaos as a source of randomness to derive an accurate kinetic equation. The difference between real Hamiltonian chaos and a conventional understanding of the laws of statistical physics can be demonstrated using the concept of Maxwell's Demon. It will be demonstrated that the Maxwell's Demon, or its equivalent, can be realised under conditions of dynamical chaos for an at least arbitrarily long period of time.

## 7.2 Fractal Traps and Maxwell's Demon

In Section 6.3, the notion of quasi-traps for Hamiltonian dynamics was introduced. It is based on the existence of singular zones in the phase space, where trajectories can be entangled for an arbitrarily long period of time without any specific characteristic time scales. The latter property indicates a possible fractal structure of the zone. The process of quasi-trapping is described in a more general way in the following paragraphs.

[2]For more information on the publications by N. Krylov, see [Kr 50]. More discussions are found in [C 79] and [Za 81].

Fig. 7.2.1. A sketch of the recurrent quasi-trap (time-wrinkle in the phase space).

When considering the trajectory of a particle in the phase space, one allows the trajectory to correspond to a finite area of chaotic motion. This area is a hypercube $\Gamma_0$ with side $R_0$ and volume $\Gamma_0 = R_0^d$, where $d$ is the phase space dimension (see Fig. 7.2.1). It is assumed that the mixing and wandering in the phase space are almost uniform everywhere except for an area inside $\Gamma_0$. Within $\Gamma_0$, a sub-area, $\Gamma_1 \subset \Gamma_0$ with volume $\Gamma_1 = R_1^d$, is then isolated. The fine structure of the process of wandering inside $\Gamma_1$ is again uniform everywhere except for the area $\Gamma_2 \subset \Gamma_1$ with volume $R_2^d$. This process of iteration can be continued until it reaches infinity. The characteristic time $T_0$ that the particle spends in the domain $\Gamma_0$ and time $T_1$ in the domain $\Gamma_1$ can be considered simultaneously. In the case when

$$\begin{aligned} \Gamma_n &= \lambda_\Gamma^n \Gamma_0 \,, \quad (\lambda_\Gamma < 1) \\ T_n &= \lambda_T^n T_0 \,, \quad (\lambda_T > 1) \end{aligned} \tag{7.2.1}$$

with scaling constants $\lambda_\Gamma, \lambda_T$, it will be called a fractal trap. A corresponding multi-fractal generalisation is possible if a distribution of the different values of $\lambda_\Gamma$ and a corresponding distribution of $\rho_T(\lambda_T)$ exist. Hence

$$\begin{aligned} \Gamma_n &= (\lambda_\Gamma^{(1)} \lambda_\Gamma^{(2)} \cdots \lambda_\Gamma^{(n)}) \Gamma_0 = \bar{\lambda}_\Gamma^n \Gamma_0 \,, \quad (n \to \infty) \\ T_n &= \lambda_T^{(1)} \lambda_T^{(2)} \cdots \lambda_T^{(n)} T_0 = \bar{\lambda}_T^n T_0 \,, \quad (n \to \infty) \end{aligned} \tag{7.2.2}$$

with

$$\bar{\lambda}_{\Gamma,T} = \int \lambda_{\Gamma,T} \rho_{\Gamma,T}(\lambda_{\Gamma,T}) d\lambda_{\Gamma,T} \,. \tag{7.2.3}$$

With Fig. 7.2.1 as a model of the fractal trap, a surprisingly simple qualitative description of its working can be provided. When the condition (7.2.1) is valid, one can express

$$n = \frac{\ln(T_n/T_0)}{\ln \lambda_T} \tag{7.2.4}$$

from the equation for $T$ and put (7.2.4) into Eq. (7.2.1) for $\Gamma_n$,

$$\frac{\Gamma_n}{\Gamma_0} = \exp(-n|\ln \lambda_\Gamma|) = \left(\frac{T_0}{T_n}\right)^{\mu_\Gamma} \,, \tag{7.2.5}$$

with the following notation:

$$\mu_\Gamma = |\ln \lambda_\Gamma| / \ln \lambda_T \,. \tag{7.2.6}$$

The construction of the trap is such that the trajectories, which spend an interval of trap time $t > T_n$, would cross the phase volume $\Gamma_n$:

$$\Gamma_n \sim 1/T_n^{\mu_\Gamma} \,. \tag{7.2.7}$$

From the general expression (6.3.12), the probability density for the trajectory to exit from the trap is derived as

$$P_R(t) \sim \psi(t; \Delta\Gamma) \sim 1/t^\gamma = 1/t^{2+\mu} \,, \tag{7.2.8}$$

where (6.4.13) and (6.4.12) are used for $\gamma$ and the definition of $\mu$, respectively:

$$\mu = |\ln \lambda_s| / \ln \lambda_T \,. \tag{7.2.9}$$

From the other side, the same trajectories filling the phase volume can be expressed through the escape probability $\psi(t;\Delta\Gamma)$ if $\Delta\Gamma$ is replaced by $\Gamma_n$:

$$\int_{T_n}^{\infty} t\psi(t;\Gamma_n)dt \sim 1/T_n^{\gamma-2} . \tag{7.2.10}$$

A comparison of (7.2.7) and (7.2.10) yields

$$\gamma = 2 + \mu_\Gamma = 2 + |\ln \lambda_\Gamma / \ln \lambda_T . \tag{7.2.11}$$

This expression coincides with (6.4.13) if $\lambda_\Gamma$ is replaced by $\lambda_s$, which is true if one takes into account that $s$ in $\lambda_s$ is the just area in the phase space. In a new way, one arrives at the expression

$$\gamma = 2 + |\ln \lambda_s| / \ln \lambda_T , \tag{7.2.12}$$

which defines the important exponent for the exit time and recurrences distribution (7.2.8).

The fractal trap can be considered a time-wrinkle in the phase space. This system of islands, with a hierarchy of sub-islands described in Section 3.5, is an example of a fractal trap. The existence of fractal traps is a remarkable property of Hamiltonian chaos and a typical feature of the phase space in real systems. Sometimes, the scaling constant $\lambda_T$ can be immensely large and immensely small for $\lambda_T$ so that one is not able to observe the phenomenon of trapping within a feasible time. But if it can be carried out in real time, as demonstrated in Section 3.5, the question arises: what kind of statistics, or even thermodynamics, are compatible with systems possessing fractal traps in the phase space? To have a clear answer to this question, one needs to consider the Maxwell's Demon.

In the *Theory of Heat* (1871), Maxwell proposed a conceptual device which enables the molecules to enter one of two equal chambers connected through a hole (Fig. 7.2.2). This device (Demon), located at the hole, should work against the thermodynamic law which causes the gas of molecules in two contacting volumes to be in equilibrium. The problem, however, gives rise to an ambiguity in its precise definition since it involves a non-physical element as part of the system. In contemporary

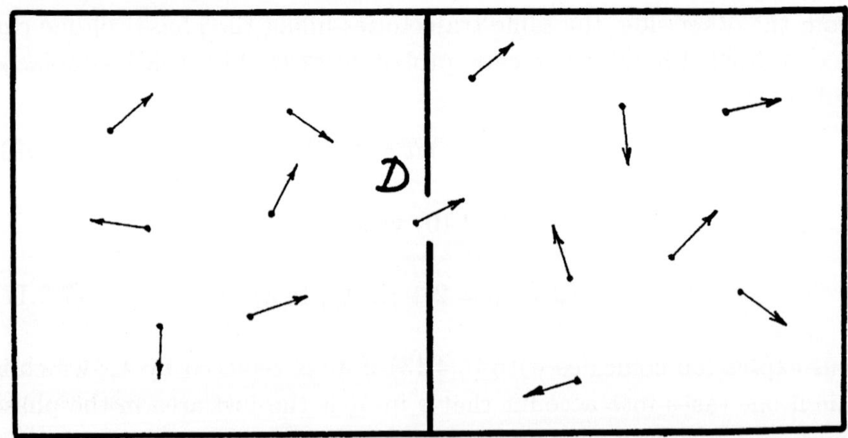


Fig. 7.2.2. A sketch of the original Maxwell's Demon design. $\mathcal{D}$ is the location of the Demon.

physics, this element is specified which enables the Maxwell's Demon to acquire a different and realistic visualisation:

(i) The Demon is able to work with information and transform the information into action (thinking device).
(ii) The Demon is a measuring device which acts upon the result of the measurement.

Both concepts give rise to fruitful physical discussions on the possibilities of computing devices, the irreversibility of computations, the natural limitations of the measurement process, and the role of quantum effects and quantum uncertainty.[3]

It is possible to propose a direction for studying the Maxwell's Demon problem based on its complete dynamical formulation and by avoiding any type of elements that cannot be formulated as equations of motion. Dynamical chaos makes it possible to formulate a new version of the Maxwell's Demon problem. In his original publication, Maxwell wrote that in statistical consideration, "... we are compelled to adopt ... the

[3]There is an excellent collection of publications [LR 90] on the Maxwell's Demon problem accompanied by comprehensive reviews.

statistical method of calculations and to abandon the strict dynamical method in which we follow every motion by calculus". Here one shall just adhere to strict dynamics.[4]

The general idea of the proposed approach to the problem can be formulated in a simple way by considering two separate dynamical systems whose trajectories have mixing properties due to the dynamical chaos. This is so that statistical equilibrium can be achieved in each system. A weak contact is then made between the two systems. A new equilibrium state is established in the coupled systems and a question thus arises: does the new equilibrium correspond to what we have commonly defined as thermodynamical equilibrium (such as equal pressures and temperatures)? The answer appears to be very problematic if elements such as fractal traps exist in at least one subsystem.

A prototype of the dynamical model can be introduced which consists of two billiard-like systems with mixing motion inside each of them and with contact being made through a hole. A point particle bounces inside the billiard with absolute elastic reflections from the billiard's walls. An example of such a system with coupled Sinai's billiards is shown in Fig. 7.2.3. Since both billiards have mixing properties, one can expect to find a stationary distribution function, such as the probability of finding a particle in one or another part of the system in the infinite time limit. For ergodic motion, the infinite time limit can be replaced by an ensemble average in the phase space of a system. Is the equilibrium in the described billiard-like system the same as thermodynamic equilibrium? In other words, can the Hamiltonian chaotic dynamics explain the origin of the laws of thermodynamics, or are some additional restrictions needed? It will be demonstrated that a negative answer will be provided to the first question using an appropriate "design" of the billiard-like system with chaotic dynamics and a fractal trap.

Actually, the main subject of this discussion is to find out what kind of random process corresponds to the dynamical chaotic motion.

---

[4]The dynamical approach to the Maxwell's Demon problem was formulated in [Za 95] and [ZE 97] based on the idea of fractal traps. They are analysed in this section.

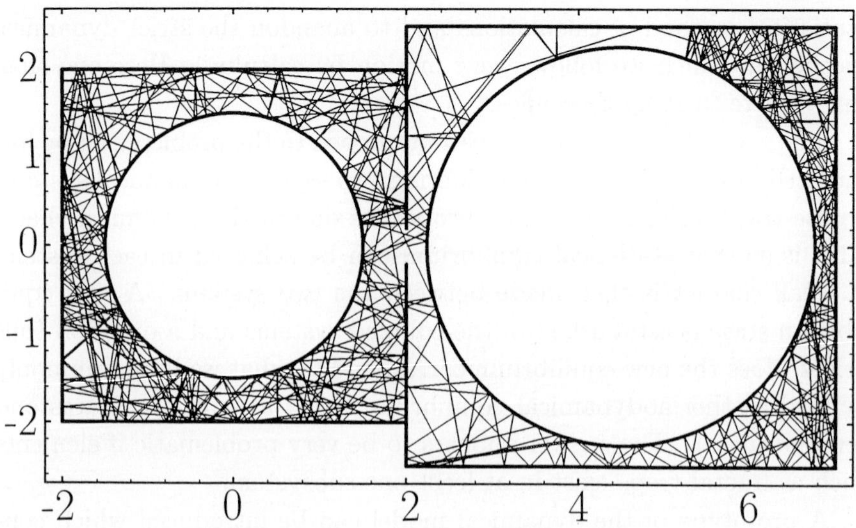


Fig. 7.2.3. Two contacting through a hole in the Sinai billiards.

It is well-known that there are serious difficulties in answering this question. For Hamiltonian systems of the general type, the motion is not ergodic. This is due to a (multi-) fractal structure of islands with regular motion. If this structure is removed from the entire phase space, the remaining phase space will correspond to the area of ergodic motion. Nevertheless, as mentioned several times before, the boundaries between the islands and the stochastic sea create singular zones that prevent the typical Hamiltonian chaotic dynamics from following a typical kinetics or regular thermodynamics.

## 7.3 Coupled Billiards

As mentioned above, in a system with two different contacting billiards having a hole in the dividing wall, a particle in a billiard is reflected elastically from the walls and scatterers. It can also pass through the hole from one billiard to another. The goal is to determine the equilibrium distributions and their moments for different characteristics of the particle trajectory. In the example given here, the system to be studied

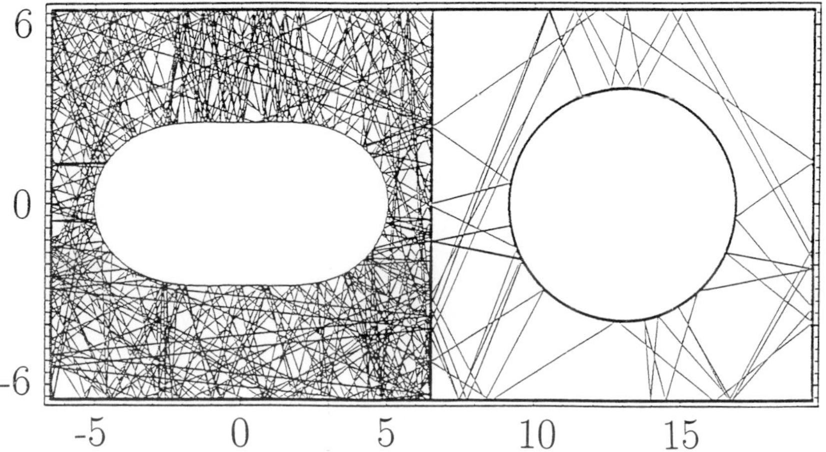


Fig. 7.3.1. Coupled Cassini-Sinai billiards. A small hole is in the centre of the partition.

consists of two subsystems: Sinai billiard (S) and Cassini billiard (C), since both have, respectively, a circle and Cassini's oval (1.6.4) as the scatterers (Fig. 7.3.1). This two-billiard system is called a CS-billiard. After a "while", one can expect a stationary distribution in the CS-billiard. It is also possible to define the distribution functions in the left and right halves of the CS-billiard to normalise these distributions and to calculate their left and right moments. The question then is: are both left and right moments similar? If the answer is negative, the result can be attributed to an absence of a thermodynamic-type equilibrium between the left and right subsystems.

Both the Sinai and Cassini billiards were introduced in Section 1.6 (see also Figs. 1.6.1 and 1.6.2). Here, more information is provided on them.[5]

The inner scatterer for the Cassini billiard has the form of a Cassini's oval (1.6.4):

$$(x^2 + y^2)^2 - 2c^2(x^2 - y^2) - (a^4 - c^4) = 0\,.$$

[5]The results follow [ZE 97].

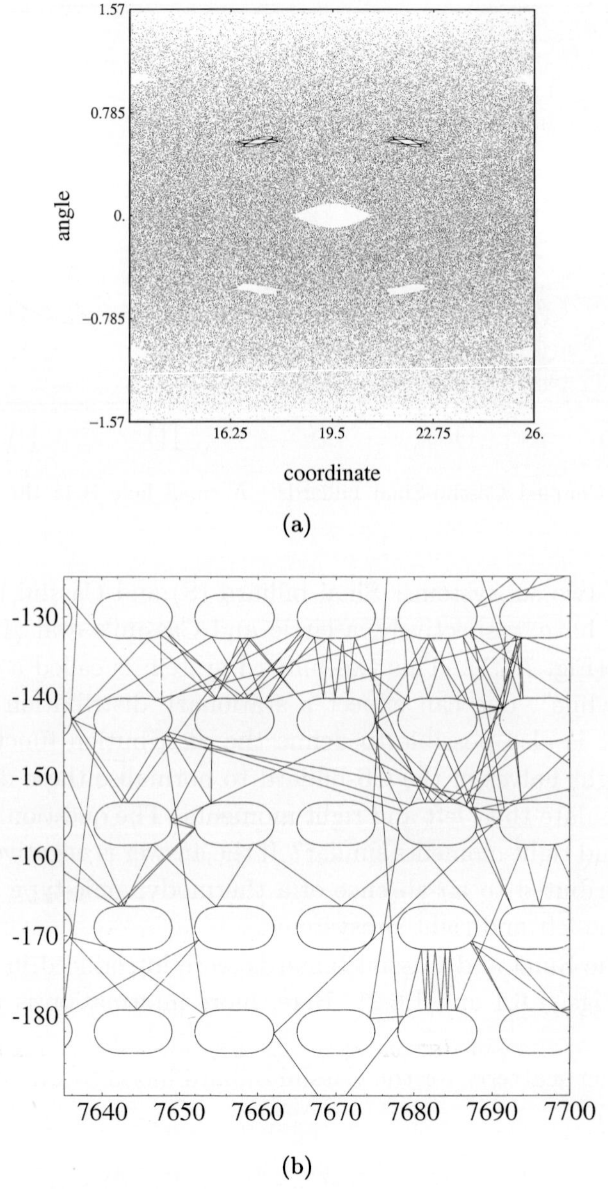


Fig. 7.3.2. Trajectories in the isolated Cassini billiard: (a) Poincaré section of the $(x, \cos^{-1} v_x)$ plane; (b) a sample of trajectory in real co-ordinate space; and (c) the same as in (b) but on a larger scale.

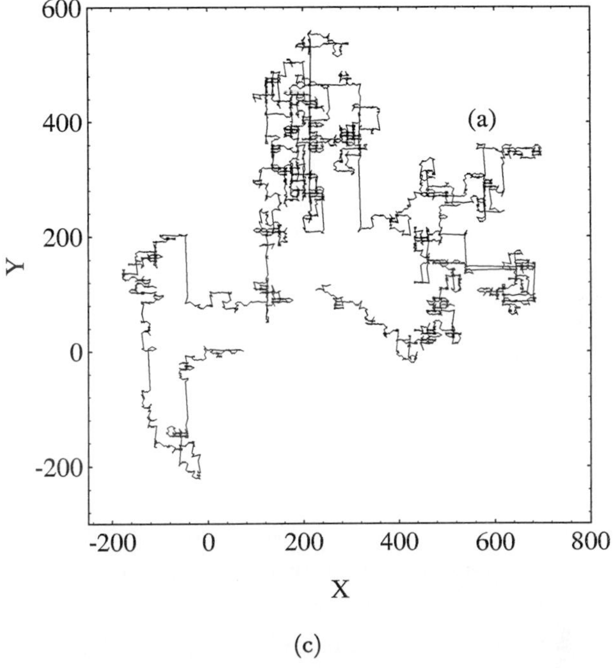


(c)

Fig. 7.3.2. (*Continued*)

The different regimes of scattering can also be created by changing the parameters $a$ and $c$. The phase space of the billiard belongs to a "regular" case, that is, islands exist in the stochastic sea if the curve $y = y(x)$ is not convex. The Poincaré section for trajectories consists of points $(x, v_x)$ or $(x, \phi)$, where $x$ is the intersection co-ordinate of a trajectory and the bottom side of the billiard and $v_x = \cos\phi$, that is, $\phi$ is the angle between the velocity vector $\mathbf{v}$ and the $x$-axis. The invariant Lebesgue measure (stationary distribution function) is non-zero on the $(x, v_x)$-plane except for the islands and zero-measure line segments which correspond to the bouncing trajectories with $v_x = \pm 1$ $(\phi = 0, \pi)$.

An example of the phase plane with islands and stochastic sea is shown in Fig. 7.3.2(a). The set of islands belongs to the fourth-order resonance and the dark strips around some islands correspond to the islands' stickiness when the trajectory spends a long time rotating around the island near its boundary. The stickiness observed in Fig. 7.3.2(a) indicates a ballistic mode regime which is better considered in the infinite phase space with periodically continued scatterers, that is, the Lorentz gas model with Cassini's oval scatterers (see Fig. 7.3.2(b)). The ballistic mode corresponds to very long segments of a trajectory which bounces between two (or more) arrays of scatterers. The same trajectory in Fig. 7.3.2 is plotted in Fig. 7.3.2(c) in infinite co-ordinate space and it reveals many flights which correspond to the trajectory stuck at the boundaries of different islands.

As discussed in Chapter 3, the occurrence of ballistic modes is a general property of Hamiltonian dynamics. They can be easily observed for some special values of parameters when an ordered set of islands is generated. In the case found in Fig. 7.3.2(a), the values are $a = 4.030952$ and $c = 3$. The corresponding alternating hierarchy of sub-islands 4-8-4-8-... is shown in Fig. 7.3.3. It was mentioned in Section 3.5 that for special values of parameters, an ordered sequence of islands with the scaling properties of the islands' space-time characteristics exists. In Section 3.5, only the value of the proliferation coefficient $q$ was considered (see Tables 3.5.1 and 3.5.2 for the standard map and the web-map) so that the number of islands for different generations follows the sequence: $q, q^2, q^3, \ldots$. In the case shown in Fig. 7.3.3, the two values of $q = 4$ and $q = 8$ alternate with each other in the proliferation coefficient as shown in Table 7.3.1.

Table 7.3.1 displays the values of the proliferation coefficient $q_k$ for the $k$-th generation, the period $T_k$ of the last invariant curve inside an island of the $k$-th generation, the area $\Delta S_k$ of an island of the $k$-th generation, and the area

$$\delta S_k = q_k \Delta S_k \tag{7.3.1}$$

of all islands of the $k$-th generation. In accordance with Section 3.5, two scaling parameters can be introduced to describe a self-similarity in the islands' hierarchy:

$$\begin{aligned} \lambda_S^{(k)} &= \delta S_k / \delta S_{k-1} \\ \lambda_T^{(k)} &= T_k / T_{k-1} \,. \end{aligned} \tag{7.3.2}$$

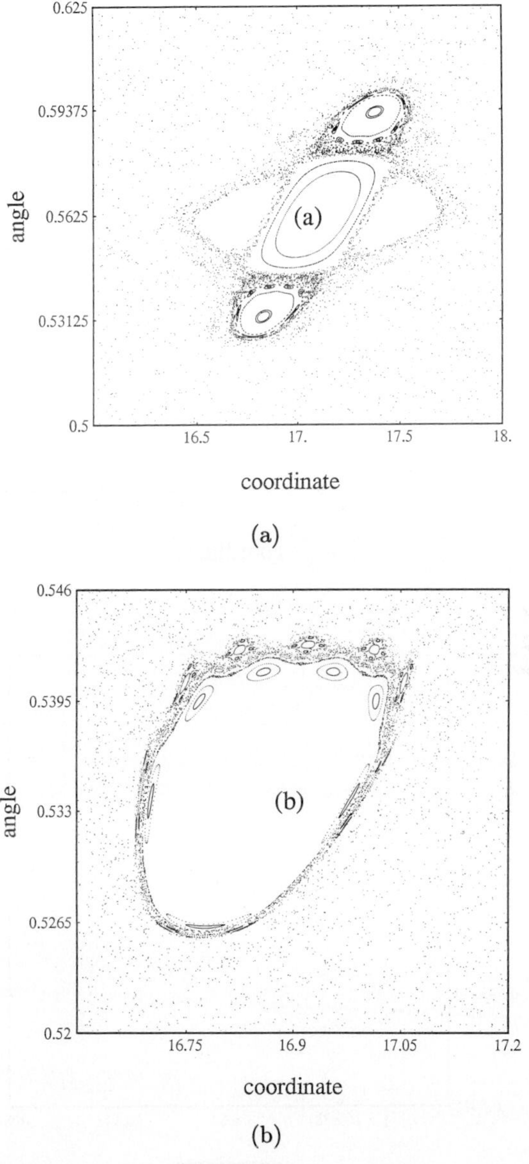


Fig. 7.3.3. An island and its vicinity for the Cassini billiard with $a = 4.030952$ and $c = 3$: (a) Poincaré plot of the initial island taken from Fig. 7.3.2(a); (b) magnification of the bottom island from (a); (c) magnification of the top left island from (b); and (d) magnification of the right island from (c).

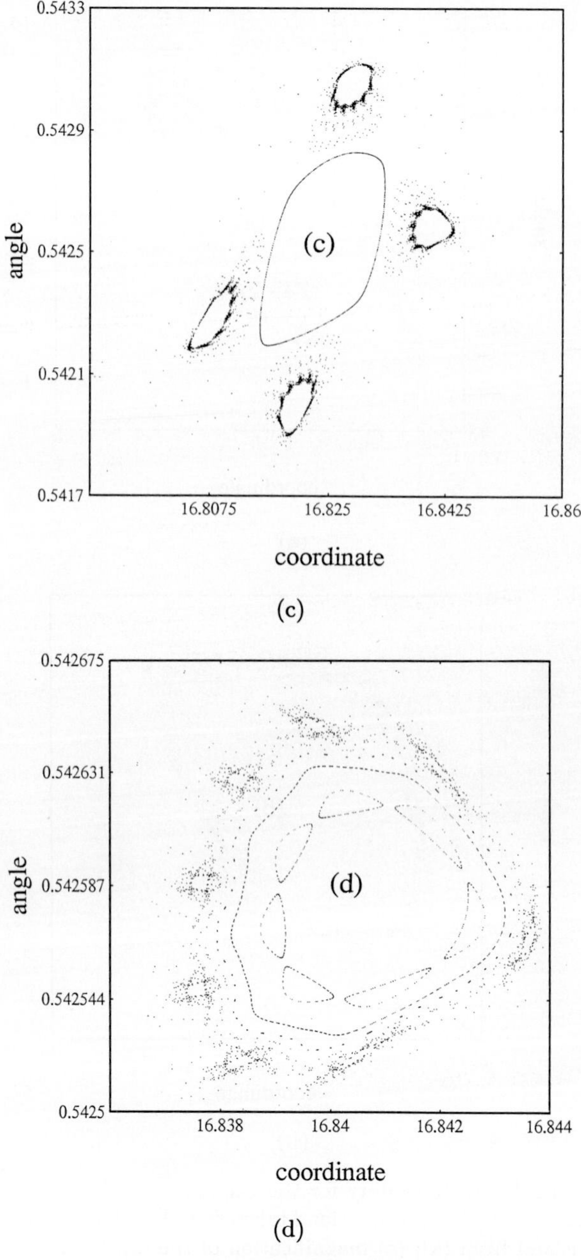


Fig. 7.3.3. (*Continued*)

Table 7.3.1. Parameters of the islands hierarchy in the sequence 2-4-8-4-8.

| $k$ | $q_k$ | $T_k$ | $T_k/T_{k-1}$ | $\Delta S_k$ | $\Delta S_k/\Delta S_{k-1}$ | $\delta S_k$ | $\delta S_k/\delta S_{k-1}$ |
|---|---|---|---|---|---|---|---|
| 0 | 2 | 16.36 | — | $1.47 \times 10^{-2}$ | — | $2.94 \times 10^{-2}$ | — |
| 1 | 4 | 118 | 7.21 | $3.96 \times 10^{-3}$ | $2.69 \times 10^{-2}$ | $3.17 \times 10^{-2}$ | 1.08 |
| 2 | 8 | 508.9 | 4.31 | $8.53 \times 10^{-6}$ | $2.15 \times 10^{-3}$ | $5.46 \times 10^{-4}$ | 0.017 |
| 3 | 4 | 3910 | 7.69 | $4.4 \times 10^{-7}$ | $5.2 \times 10^{-2}$ | $1.1 \times 10^{-4}$ | 0.21 |
| 4 | 8 | 15740 | 4.02 | $0.96 \times 10^{-10}$ | $2.2 \times 10^{-3}$ | $2.0 \times 10^{-6}$ | 0.018 |

For the constant value $q_k = q$ $(\forall k \geq 1)$, a simulation confirms the existence of constant values of scaling parameters

$$\lambda_S^{(k)} = \lambda_S\,, \quad \lambda_T^{(k)} = \lambda_T\,, \quad (\forall k \geq 1) \tag{7.3.3}$$

for the web-map and standard map (see Section 3.5). In the case of Fig. 7.3.3, the Cassini billiard is faced with a new situation. It has the two values of $\lambda_S^{(1,2)}$ and $\lambda_T^{(1,2)}$ which are found in Table 7.3.1:

$$\begin{aligned} &\lambda_T^{(1)} \sim 7.4\,, \qquad \lambda_T^{(2)} \sim 4.2\,, \\ &\lambda_S^{(1)} \sim 0.017\,, \quad \lambda_S^{(2)} \sim 0.21\,, \end{aligned} \tag{7.3.4}$$

where the mean values are taken and $\delta S_1/\delta S_0$ is skipped since it does not correspond to the set $(q_0 \neq 8)$. These values of the scaling parameters can be used to analyse the Poincaré recurrences distribution function in kinetics which will be discussed in Chapter 11.

## 7.4 Contacted Cassini-Sinai Billiards

One now considers the model in Fig. 7.3 which consists of both the Cassini (left chamber) and Sinai (right chamber) billiards contacting through a hole in their dividing wall. Using a trajectory and the time instants $\left\{t_j^{(C)}\right\}$ and $\left\{t_j^{(S)}\right\}$ when a particle leaves $C$ (Cassini chamber) or $S$ (Sinai chamber), respectively, the sequences

$$\begin{aligned} \left\{\tau_{j+1}^{(C)}\right\} &= \left\{t_{j+1}^{(C)} - t_j^{(S)}\right\} \\ \left\{\tau_{j+1}^{(S)}\right\} &= \left\{t_{j+1}^{(S)} - t_j^{(C)}\right\} \end{aligned} \tag{7.4.1}$$

can be related to the residence times for $C$ and $S$ when the systems are in contact. One can also say that sequences (7.4.1) are recurrence times for the domain $\Delta$ covered by the hole, and that $\left\{\tau_j^{(C)}\right\}$ and $\left\{\tau_j^{(S)}\right\}$ are sets of the left or right recurrences, respectively. When there is no hole, $t_j^{(S)}$ should be replaced by $t_j^{(C)}$ and $t_j^{(C)}$ by $t_j^{(S)}$ in (7.4.1). The Poincaré recurrences for the $C$ and $S$ billiards can therefore be obtained independently.[5]

A corresponding simulation was performed for a single trajectory during $t \cong 10^{11}$ which corresponds to about $10^{10}$ crossings of a billiard (a chamber). The size of the hole was 0.2 and that of one side was 13. The phase volumes of both billiards were the same, with an accuracy of up to $10^{-3}$.

The results of the probability distribution densities, $P(t, \Delta)$, for Poincaré recurrences in the domain $\Delta$ in the case of independent

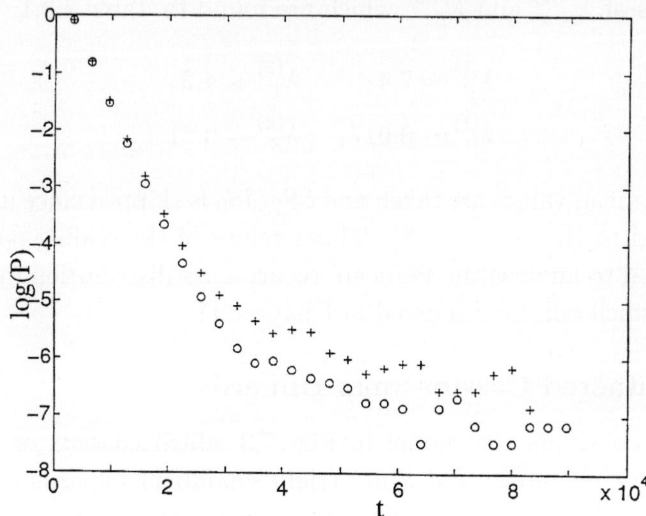


Fig. 7.4.1. Distribution of the Poincaré recurrences for the isolated Cassini and Sinai billiards. Crosses denote the Cassini billiard and circles the Sinai billiard.

(isolated) $C$ and $S$ billiards are shown in Fig. 7.4.1. The probability follows the Poissonian law (see Section 6.2) of up to $t_0 \sim 2 \cdot 10^4$ at the same mean time, $\langle\tau\rangle$, which corresponds to the results mentioned in Sections 6.1 and 6.2. The value of $t_0$ corresponds to the chosen domain $\Delta$. A check on the behaviour obtained in Fig. 7.4.1 shows that it does not depend on the size and location of the domain $\Delta$, but the crossover time $t_0$ does.

For $t > 2 \cdot 10^4$, significant deviations from the Poissonian law occur and long tales have been observed. The difference between these two distributions is evident and can be expressed more clearly using the high moments of $P(t, \Delta)$:

$$\langle\tau^m\rangle = \int_0^{t_{\max}} t^m P(t, \Delta) dt\,. \tag{7.4.2}$$

Figure 7.4.2 presents the corresponding moments up to $m = 10$ normalised to $\langle t\rangle$ for $C$ and $S$ with $t_{\max} \sim 10^{11}$. Beginning with $m = 5$,

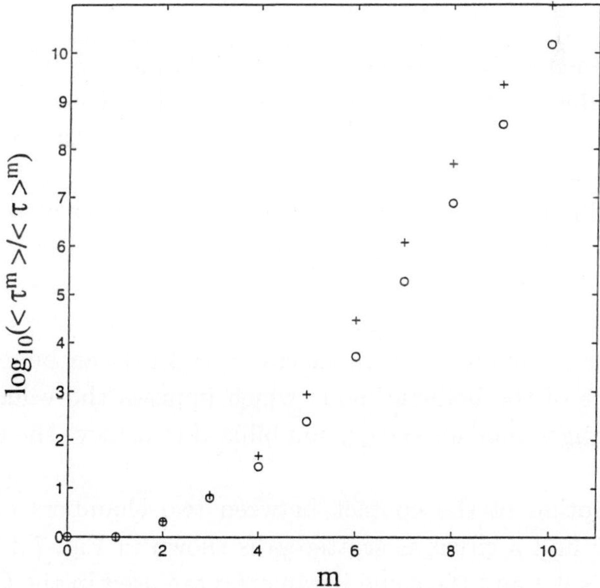


Fig. 7.4.2. Values of moments for the distributions are shown in Fig. 7.4.1 (using the same notations).

the difference approaches the value of about one order of a magnitude. Nevertheless, what is important is not only the difference between the recurrences distribution functions and their high moments, but also the existence of a difference between the first moments which are

$$\begin{aligned} \langle t^{(C)} \rangle &= 1.97 \cdot 10^3 \,, \quad \langle t^{(S)} \rangle = 1.92 \cdot 10^3 \\ \langle \Delta t^{(CS)} \rangle &= \langle t^{(C)} \rangle - \langle t^{(S)} \rangle = 0.05 \cdot 10^3 \,. \end{aligned} \tag{7.4.3}$$

This is consistent with the level of resolution in computations, that is, $\langle \Delta t^{(CS)} \rangle$ is larger than the value $10^{-3} \langle t^{(C,S)} \rangle$ which is the accuracy of the phase volume evaluation. The difference $\langle \Delta t^{(CS)} \rangle \neq 0$ occurs because the distribution of recurrences is still not stationary despite the very long computation time. Moreover, the tail influence is still sensitive to the value $\langle t^{(C,S)} \rangle$.

The exponent $\gamma$ of the power-tail in the recurrences distribution

$$P(t) \sim \text{const}/t^{\gamma} \tag{7.4.4}$$

can be obtained with a longer computation of $t_{\max} = 2 \cdot 10^{11}$ (Fig. 7.4.3). Hence one derives $\gamma = 3.02$ for the Sinai billiard and $\gamma = 3.15$ for the Cassini billiard. These values are consistent with the theoretical prediction $\gamma = 3$.[6]

The origin of the value $\gamma = 3$ is due to the presence of non-scattered bouncing trajectories and the corresponding singularity of the phase space. The same type of singularity exists on the surface of a scatterer independently, and one can expect a low universality of $\gamma = 3$. In fact, for the Cassini billiard, a small excess ($\gamma = 3.15$) can be explained by the influence of the ballistic mode which imposes the value $\gamma > 3$. It requires a longer time for the Cassini billiard to achieve the asymptotics $\gamma = 3$.

A description of the contact between two chambers containing a Cassini oval and a circle as scatterers is shown in Fig. 7.3.1. The size of the hole is 0.2 and the same parameters are used in the Cassini oval, shown in Fig. 7.3.2, in order to increase its anomalous (non-Gaussian)

[6] For the value $\gamma = 3$ in the Sinai billiard with infinite horizon, see [Bl 92].

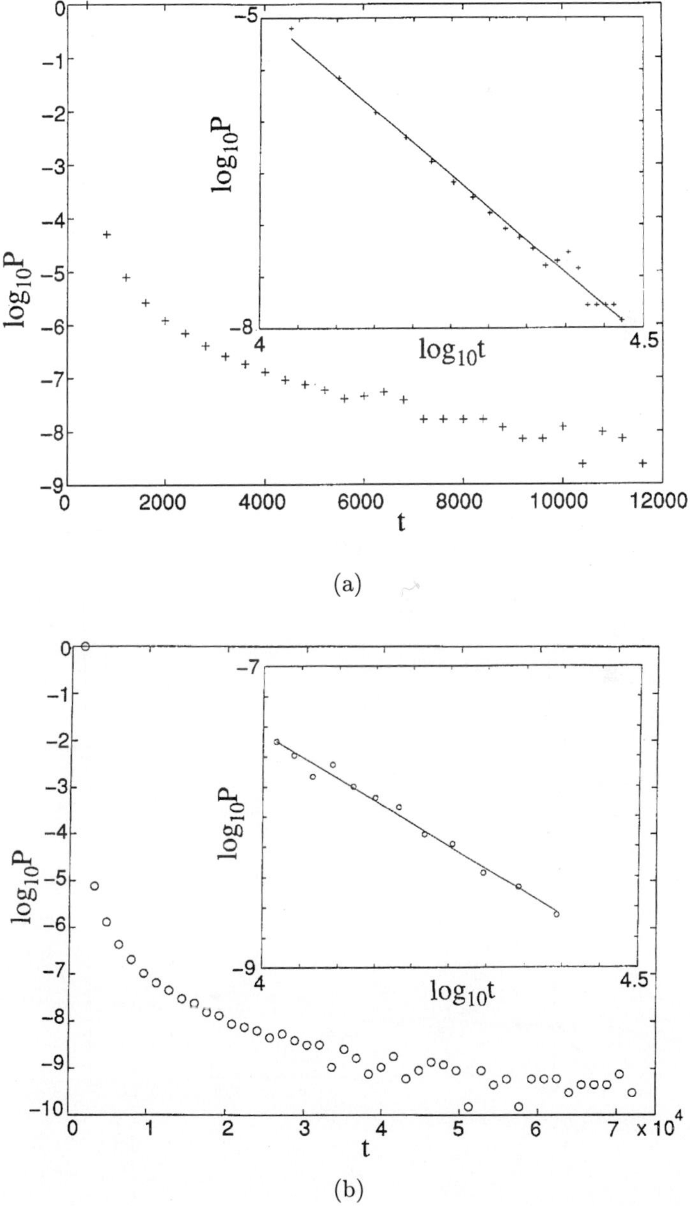


Fig. 7.4.3. The same as in Fig. 7.4.1 for isolated (a) Cassini and (b) Sinai billiards but for a longer computational time.

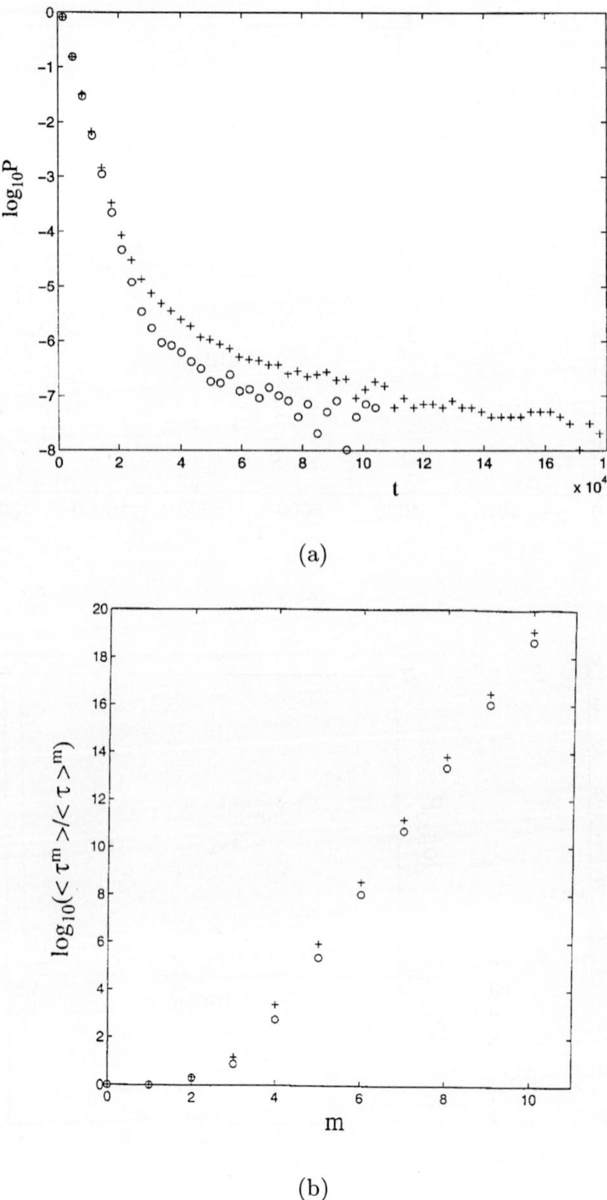


Fig. 7.4.4. Coupled Cassini-Sinai billiards: (a) distribution of Poincaré recurrences in Cassini (crosses) and Sinai (circles) parts; and (b) moments of the distributions shown in (a).

kinetics. The parameters of the Sinai billiard were adjusted so that both chambers have equal phase volumes. The corresponding distributions, $P_C(t)$ and $P_S(t)$, for the residence times defined in (7.4.1) for both the left (Cassini billiard) and right (Sinai billiard) chambers are presented in Fig. 7.4.4(a). The corresponding moments, $\langle t_C^m \rangle$ and $\langle t_S^m \rangle$, for the distributions, $P_C(t)$ and $P_S(t)$, are displayed in Fig. 7.4.4(b).

One can conclude from the results presented in Fig. 7.4.4 that there is no equilibrium in the usual thermodynamic sense, at least during the observation time $t_{\max} \sim 10^{11}$ or $\sim 10^{10}$ characteristic periods. It was mentioned above that the distributions $P_C(t)$ and $P_S(t)$ are actually the Poincaré cycle distributions and they should not depend on the location of a volume, $\Delta\Gamma$, of the observation in the case of macroscopic equilibrium. In this sense, the situation described does not correspond to an equlibrium since the distribution functions and their moments are significantly different for both chambers. One can also clarify the absence of a fast relaxation process which can establish equilibrium in a finite time.

The absence of the equilibrium in the usual thermodynamic sense during an astronomical time can also be obtained for the non-equal Sinai contact of two billiards (Fig. 7.2.3). Although the phase volumes are equal, a difference in geometry leads to a difference in the distribution functions of the recurrences. In the absence of power-like tails, one can expect a relaxation in the equilibrium distribution after a certain time interval. It has not happened in the cases considered so far, even for a contact between two different Sinai billiards, because the tails of their distribution functions produce extremely long-lived fluctuations that do not dissipate over a finite time.

The models of a two-billiard contact are very demonstrative and are fairly easy to simulate. At the same time, they do not possess too strong an effect on the absence of the thermodynamic equilibrium. This is because free-bouncing ballistic trajectories do not induce a strong singularity in the phase space. One can expect a stronger effect in the presence of a singular zone related to the accelerator mode described in Section 3.2. There could be a much stronger difference in the distribution function depending on where and how far from the singular zone the domain of observation is taken. The difference will, of course, dissipate over time. However, the time required exceeds any reasonable value and cannot be included in any physical considerations. Hence there is a need for another type of thermodynamics which would include a possibility of long-lasting fluctuations. The dynamical model described, that is, the Maxwell's Demon, works because the equilibrium conditions cannot be formulated on the basis of the canonical laws of thermodynamics.

## Conclusions

1. The problem of the foundation of statistical physics lies in the lack of a clear formulation of the conditions which make it possible to derive statistical laws and irreversibility from the first principles, that is, from the reversible Hamiltonian equations. The theory of chaotic dynamics plays an important role in solving the problem which can be reformulated in the following way: is the condition of the occurrence of chaos sufficient to obtain a typical statistical irreversibility?

2. The distribution of the Poincaré recurrences plays a crucial role in an understanding of the general properties of chaotic dynamics. Hamiltonian systems do not have traps or sources. Nevertheless, Hamiltonian dynamics permit the existence of quasi-traps with an asymptotic power-wise distribution at recurrence time $\tau$:

   $$P_R(\tau) \sim 1/\tau^\gamma \,.$$

   In some cases of space-time self-similarity, the exponent $\gamma$ can be expressed as

   $$\gamma = 2 + |\ln \lambda_\Gamma| / \ln \lambda_T \,,$$

   where $\lambda_\Gamma$ and $\lambda_T$ are the scaling characteristics of the fractal phase space-time properties in the quasi-trap.

3. The existence of quasi-traps creates a situation similar to the Maxwell's Demon. This means that chaotic dynamics exhibit some memory-type features which have to be suppressed in order to derive the laws of thermodynamics.

# COLOUR PLATES (C.1–C.8)

In all the colour plates, except for C.4, there is one trajectory plot (per plate) with a periodic (after each $10^3$ steps) change in a colour. Plates C.1–C.3 are related to the fivefold symmetry stochastic web with different time length. Plate C.4 represents a skeleton for the fivefold symmetric stochastic web. Different colours correspond to different energy intervals. Plates C.5–C.8 are related to the $(u, v)$-projection for a trajectory of the four-dimensional map (9.2.13). They correspond to small $K$ (of order 0.1) and very small $\beta^2$ (of order $10^{-4} \div 10^{-5}$) and different symmetries. One can observe very rapid diffusion and flights as well as a quasi-trap (C.8) for a time of order $10^6$.

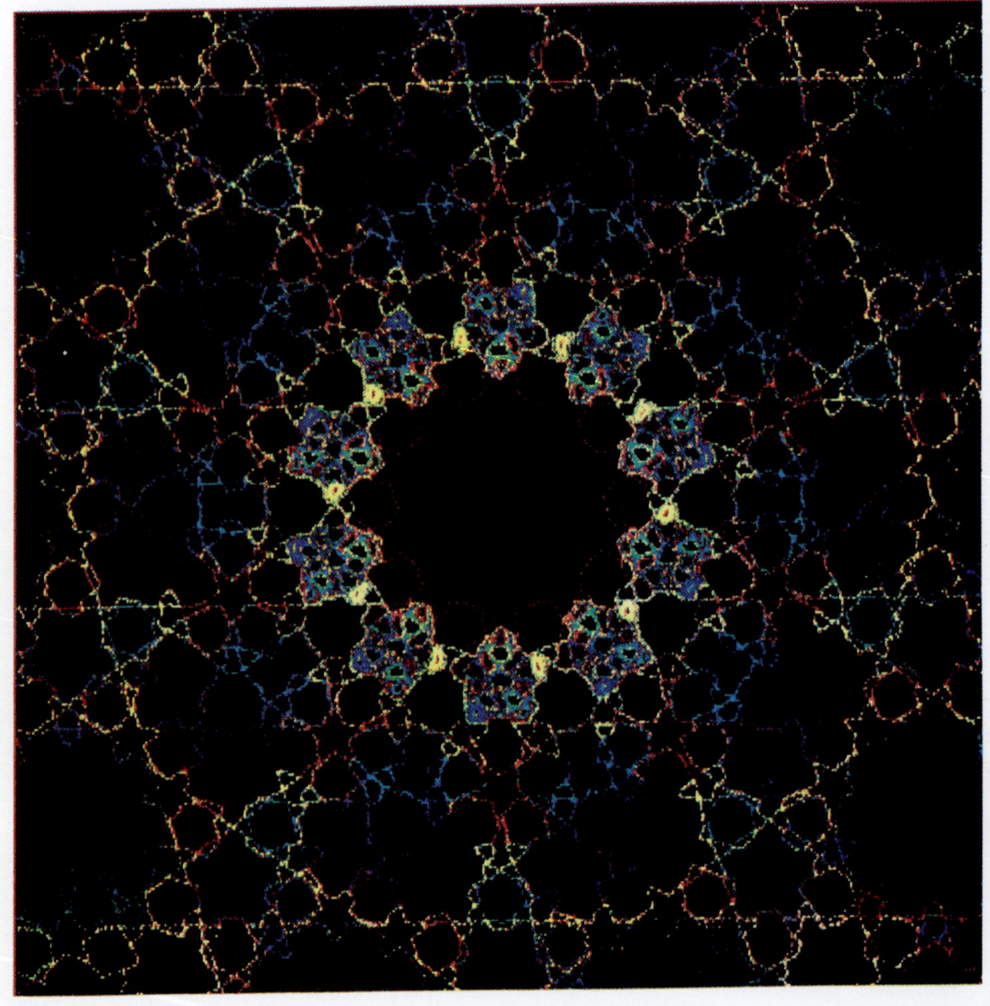

Plate C.1

Plate C.2

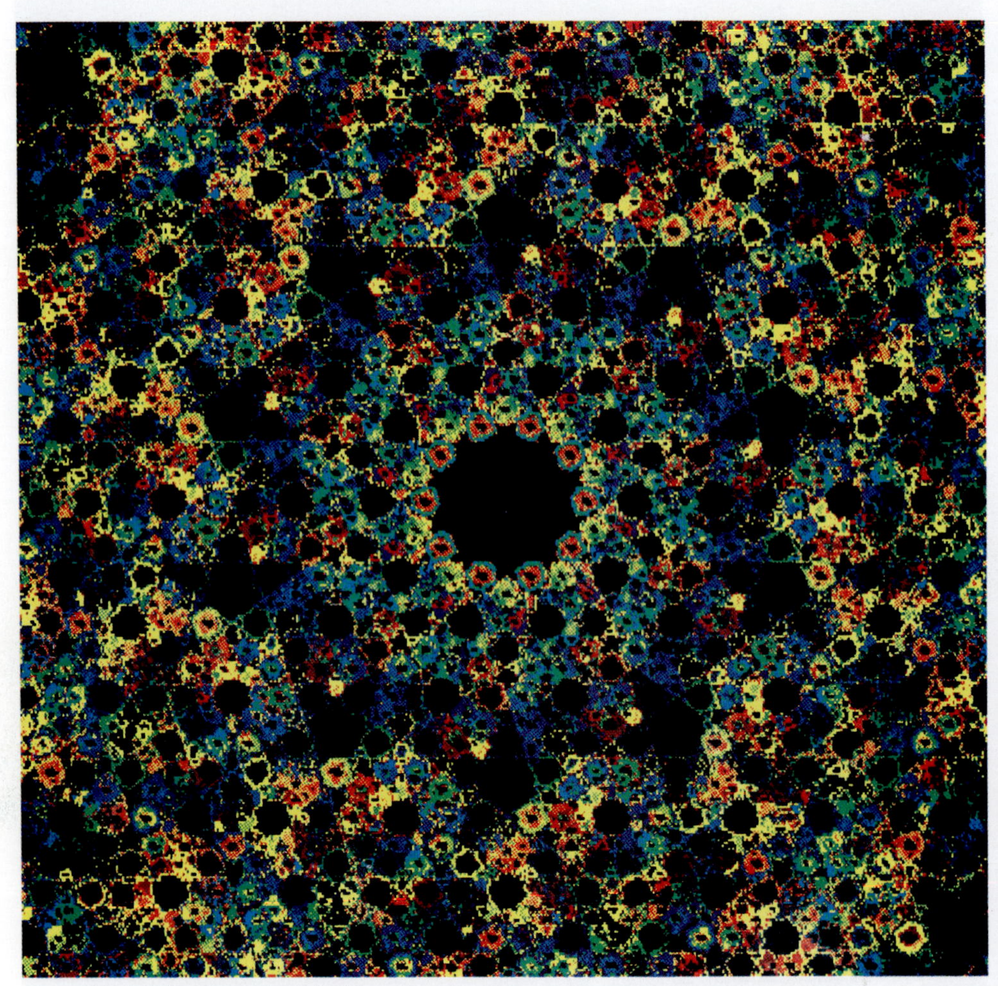

Plate C.3

Plate C.4

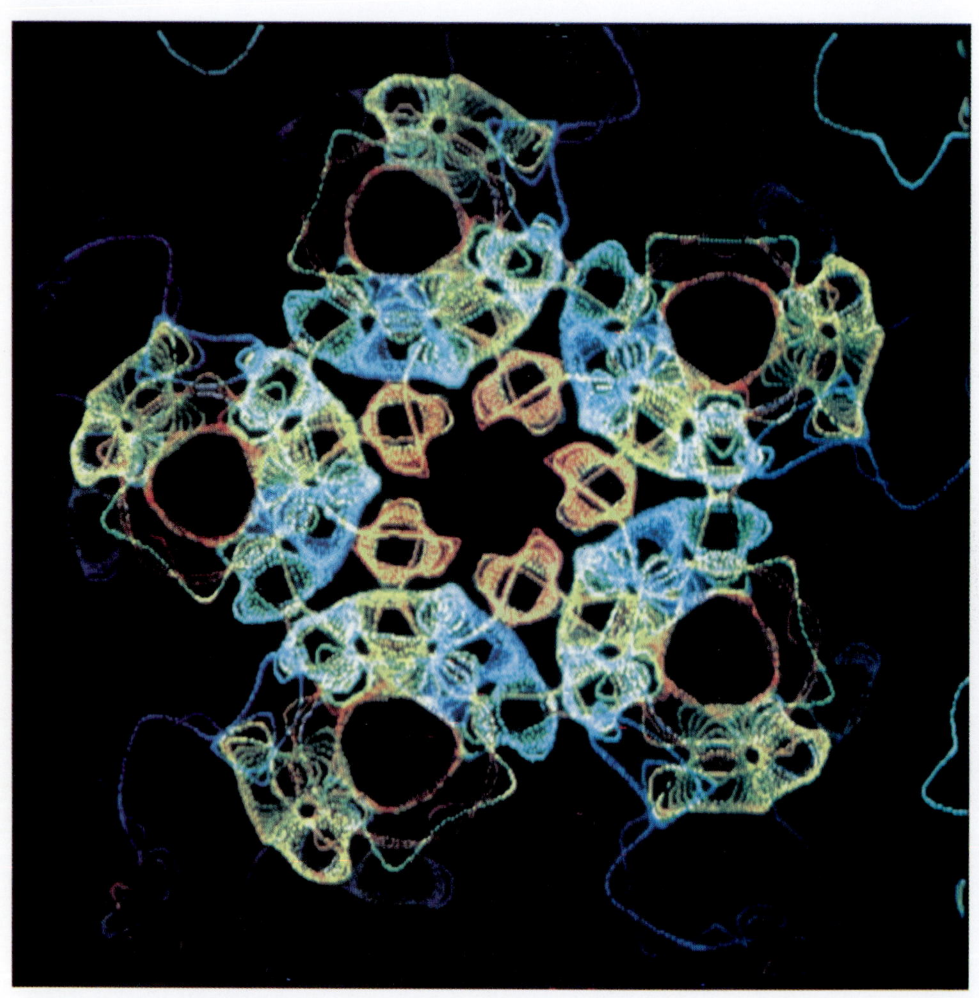

Plate C.5

Plate C.6

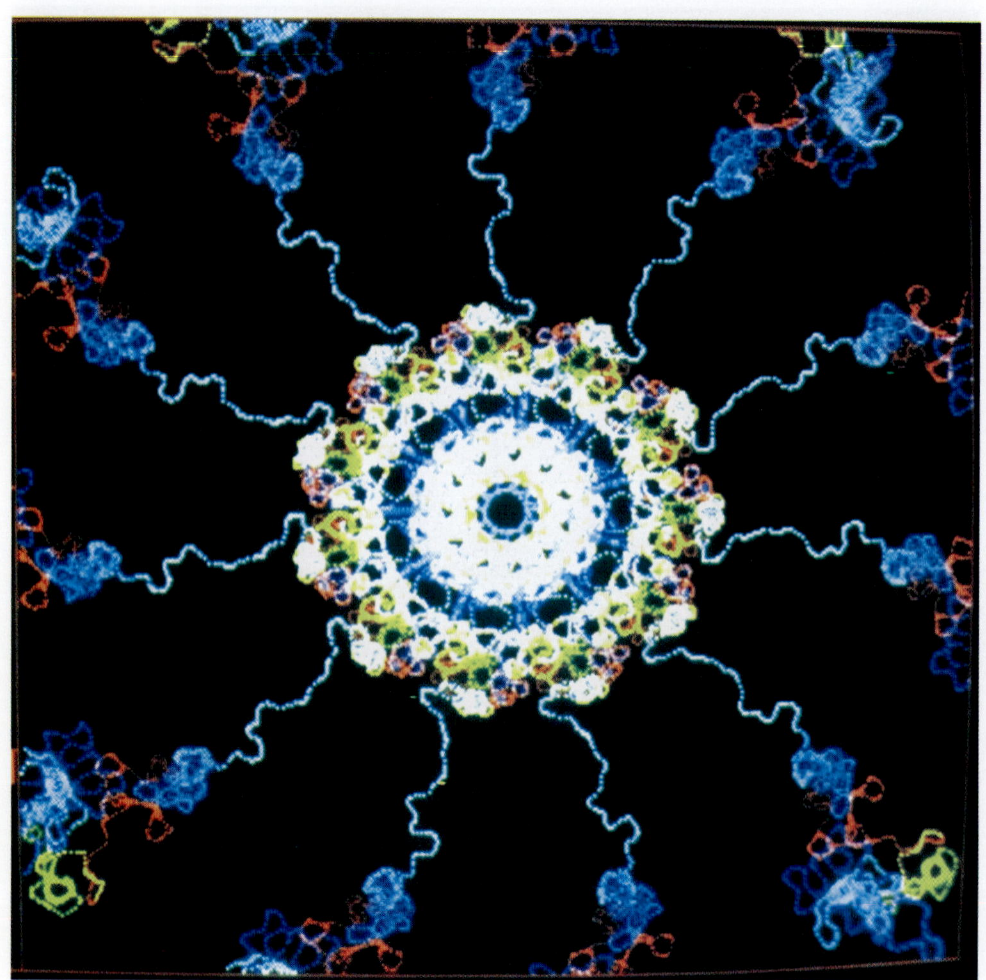

Plate C.7

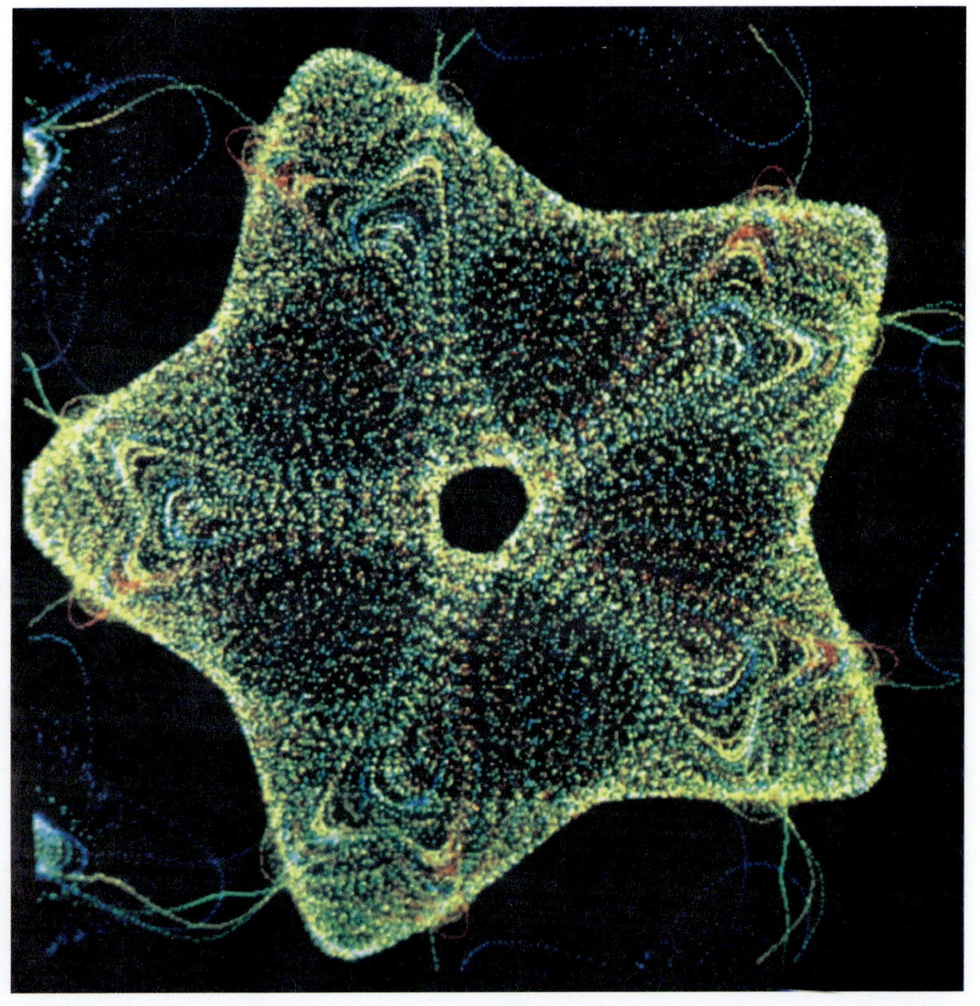

Plate C.8

# *Chapter 8*

# CHAOS AND SYMMETRY

## 8.1 Stochastic Webs

A system is close to an integrable one if its Hamiltonian can be presented as

$$H = H_0(I) + \epsilon V(I, \vartheta), \tag{8.1.1}$$

where $I = (I_1, \ldots, I_N)$ and $\vartheta = (\vartheta_1, \ldots, \vartheta_N)$ are the $N$-dimensional vectors of action and angle, respectively, and $\epsilon$ is a small parameter of perturbation. For $\epsilon = 0$, the system is described by its Hamiltonian $H_0$ and possesses $N$ first commuting integrals of motion (actions). The motion is performed along an $N$-dimensional torus defined by a specific value of $I$. In the set of tori, some of them are singular and correspond to separatrices or hypersurfaces with self-intersections. In two-dimensional phase space, separatrices are special trajectories that pass through the saddle point(s). The set of separatrices can form a connected net in the phase space (see Fig. 8.1.1) where a particle can move along. In fact, the likelihood of a particle moving along a separatrix is minimal since the time taken to reach a saddle point is infinite and it is not able to pass through the first saddle. Inside the meshes of the net, the motion is finite. This leads one to conclude that the general type of integrable motion is partitioned by a net of separatrices in an integrable case, in addition to the Liouville-Arnold theorem on integrability.[1]

[1] A rigorous formulation of the integrability conditions is very important. This can be found in [A 89]. A non-trivial example of the violation of the conditions formulated in [A 89] is found in [RB 81].

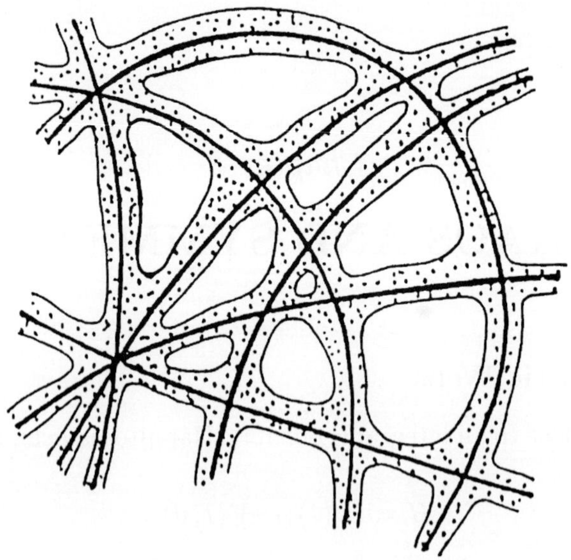

Fig. 8.1.1. Stochastic web formation.

The influence of a small perturbation in (8.1.1) does not change significantly the motion inside the meshes of the net. However, it causes a crucial replacement of the separatrices by stochastic layers. A connected net of channels of finite width along which an effective particle transport can be performed is therefore created. The set of channels is called a stochastic web (see Fig. 8.1.1).

Different kinds of finite or infinite stochastic webs are possible, including those with or without a symmetry. It is convenient to introduce a notion of the web's skeleton since the width of the web tends to zero if $\epsilon \to 0$. It is possible for the skeleton of the web not to coincide with the separatrices net for $\epsilon = 0$, and the web can only occur when there is perturbation. Such a case takes place for the periodically kicked oscillator problem (1.3.1),

$$H = \frac{1}{2}(\dot{x}^2 + \omega_0^2 x^2) - \frac{\omega_0}{T} K \cos x \sum_{m=-\infty}^{\infty} \delta\left(\frac{t}{T} - n\right), \qquad (8.1.2)$$

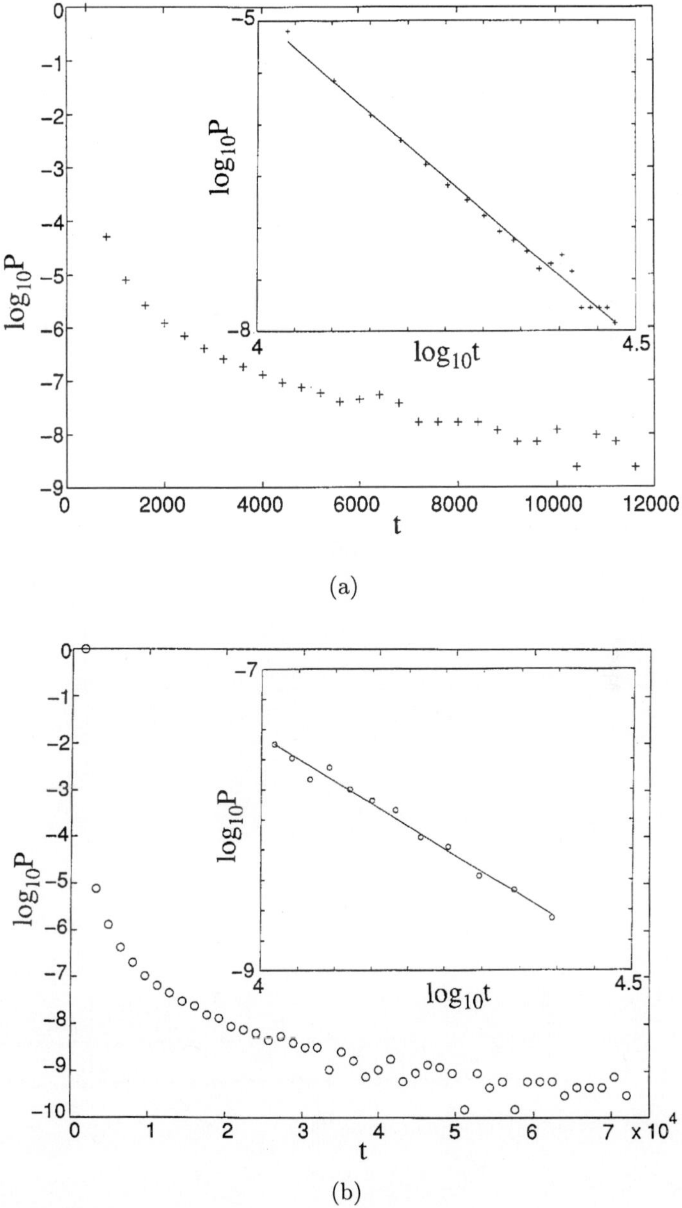


Fig. 7.4.3. The same as in Fig. 7.4.1 for isolated (a) Cassini and (b) Sinai billiards but for a longer computational time.

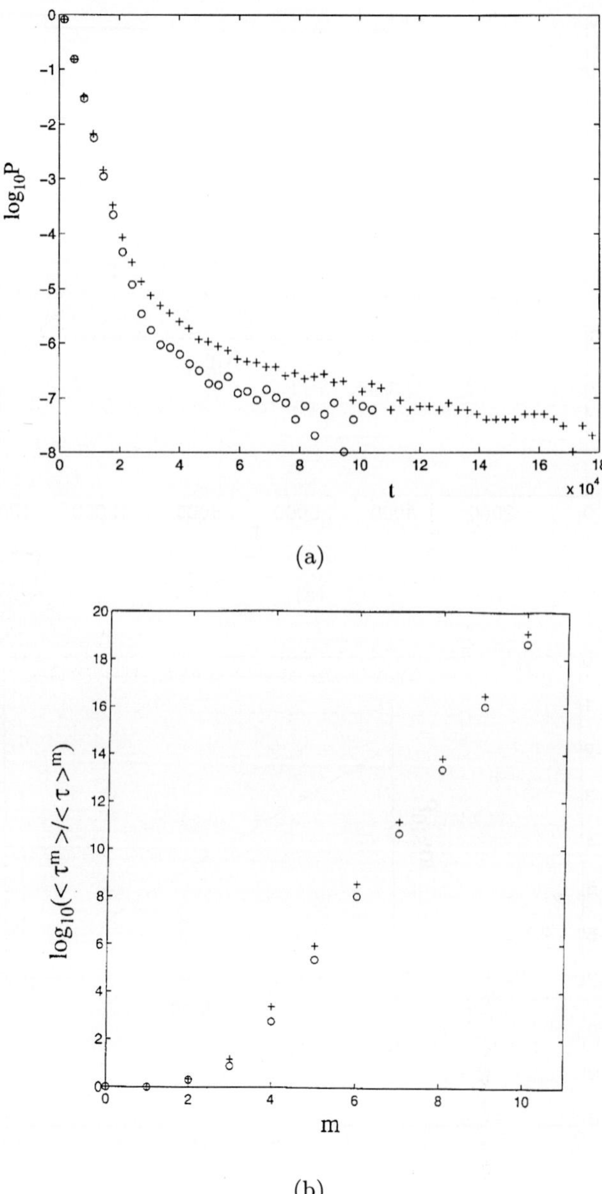


Fig. 7.4.4. Coupled Cassini-Sinai billiards: (a) distribution of Poincaré recurrences in Cassini (crosses) and Sinai (circles) parts; and (b) moments of the distributions shown in (a).

kinetics. The parameters of the Sinai billiard were adjusted so that both chambers have equal phase volumes. The corresponding distributions, $P_C(t)$ and $P_S(t)$, for the residence times defined in (7.4.1) for both the left (Cassini billiard) and right (Sinai billiard) chambers are presented in Fig. 7.4.4(a). The corresponding moments, $\langle t_C^m \rangle$ and $\langle t_S^m \rangle$, for the distributions, $P_C(t)$ and $P_S(t)$, are displayed in Fig. 7.4.4(b).

One can conclude from the results presented in Fig. 7.4.4 that there is no equilibrium in the usual thermodynamic sense, at least during the observation time $t_{\max} \sim 10^{11}$ or $\sim 10^{10}$ characteristic periods. It was mentioned above that the distributions $P_C(t)$ and $P_S(t)$ are actually the Poincaré cycle distributions and they should not depend on the location of a volume, $\Delta\Gamma$, of the observation in the case of macroscopic equilibrium. In this sense, the situation described does not correspond to an equlibrium since the distribution functions and their moments are significantly different for both chambers. One can also clarify the absence of a fast relaxation process which can establish equilibrium in a finite time.

The absence of the equilibrium in the usual thermodynamic sense during an astronomical time can also be obtained for the non-equal Sinai contact of two billiards (Fig. 7.2.3). Although the phase volumes are equal, a difference in geometry leads to a difference in the distribution functions of the recurrences. In the absence of power-like tails, one can expect a relaxation in the equilibrium distribution after a certain time interval. It has not happened in the cases considered so far, even for a contact between two different Sinai billiards, because the tails of their distribution functions produce extremely long-lived fluctuations that do not dissipate over a finite time.

The models of a two-billiard contact are very demonstrative and are fairly easy to simulate. At the same time, they do not possess too strong an effect on the absence of the thermodynamic equilibrium. This is because free-bouncing ballistic trajectories do not induce a strong singularity in the phase space. One can expect a stronger effect in the presence of a singular zone related to the accelerator mode described in Section 3.2. There could be a much stronger difference in the distribution function depending on where and how far from the singular zone the domain of observation is taken. The difference will, of course, dissipate over time. However, the time required exceeds any reasonable value and cannot be included in any physical considerations. Hence there is a need for another type of thermodynamics which would include a possibility of long-lasting fluctuations. The dynamical model described, that is, the Maxwell's Demon, works because the equilibrium conditions cannot be formulated on the basis of the canonical laws of thermodynamics.

## Conclusions

1. The problem of the foundation of statistical physics lies in the lack of a clear formulation of the conditions which make it possible to derive statistical laws and irreversibility from the first principles, that is, from the reversible Hamiltonian equations. The theory of chaotic dynamics plays an important role in solving the problem which can be reformulated in the following way: is the condition of the occurrence of chaos sufficient to obtain a typical statistical irreversibility?

2. The distribution of the Poincaré recurrences plays a crucial role in an understanding of the general properties of chaotic dynamics. Hamiltonian systems do not have traps or sources. Nevertheless, Hamiltonian dynamics permit the existence of quasi-traps with an asymptotic power-wise distribution at recurrence time $\tau$:

$$P_R(\tau) \sim 1/\tau^{\gamma} \, .$$

   In some cases of space-time self-similarity, the exponent $\gamma$ can be expressed as

$$\gamma = 2 + |\ln \lambda_\Gamma| / \ln \lambda_T \, ,$$

   where $\lambda_\Gamma$ and $\lambda_T$ are the scaling characteristics of the fractal phase space-time properties in the quasi-trap.

3. The existence of quasi-traps creates a situation similar to the Maxwell's Demon. This means that chaotic dynamics exhibit some memory-type features which have to be suppressed in order to derive the laws of thermodynamics.

# COLOUR PLATES (C.1–C.8)

In all the colour plates, except for C.4, there is one trajectory plot (per plate) with a periodic (after each $10^3$ steps) change in a colour. Plates C.1–C.3 are related to the fivefold symmetry stochastic web with different time length. Plate C.4 represents a skeleton for the fivefold symmetric stochastic web. Different colours correspond to different energy intervals. Plates C.5–C.8 are related to the $(u, v)$-projection for a trajectory of the four-dimensional map (9.2.13). They correspond to small $K$ (of order 0.1) and very small $\beta^2$ (of order $10^{-4} \div 10^{-5}$) and different symmetries. One can observe very rapid diffusion and flights as well as a quasi-trap (C.8) for a time of order $10^6$.

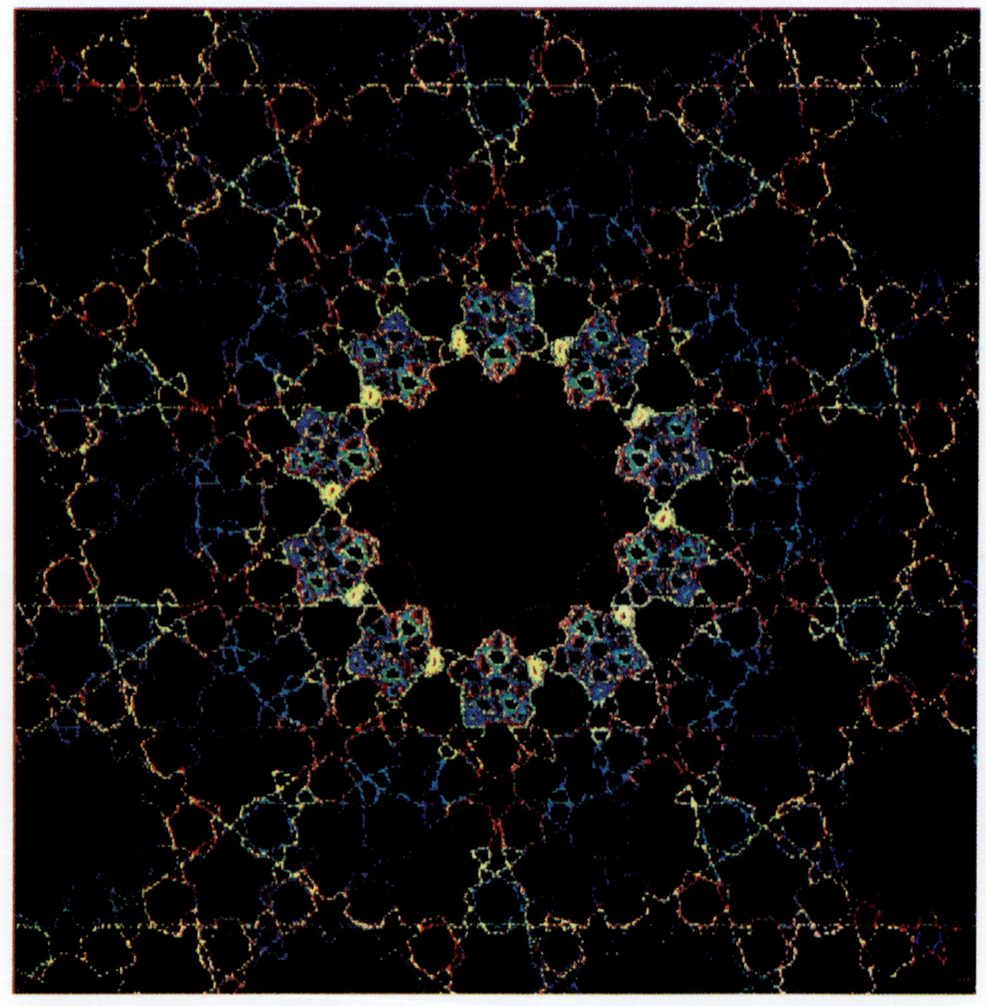

Plate C.1

Plate C.2

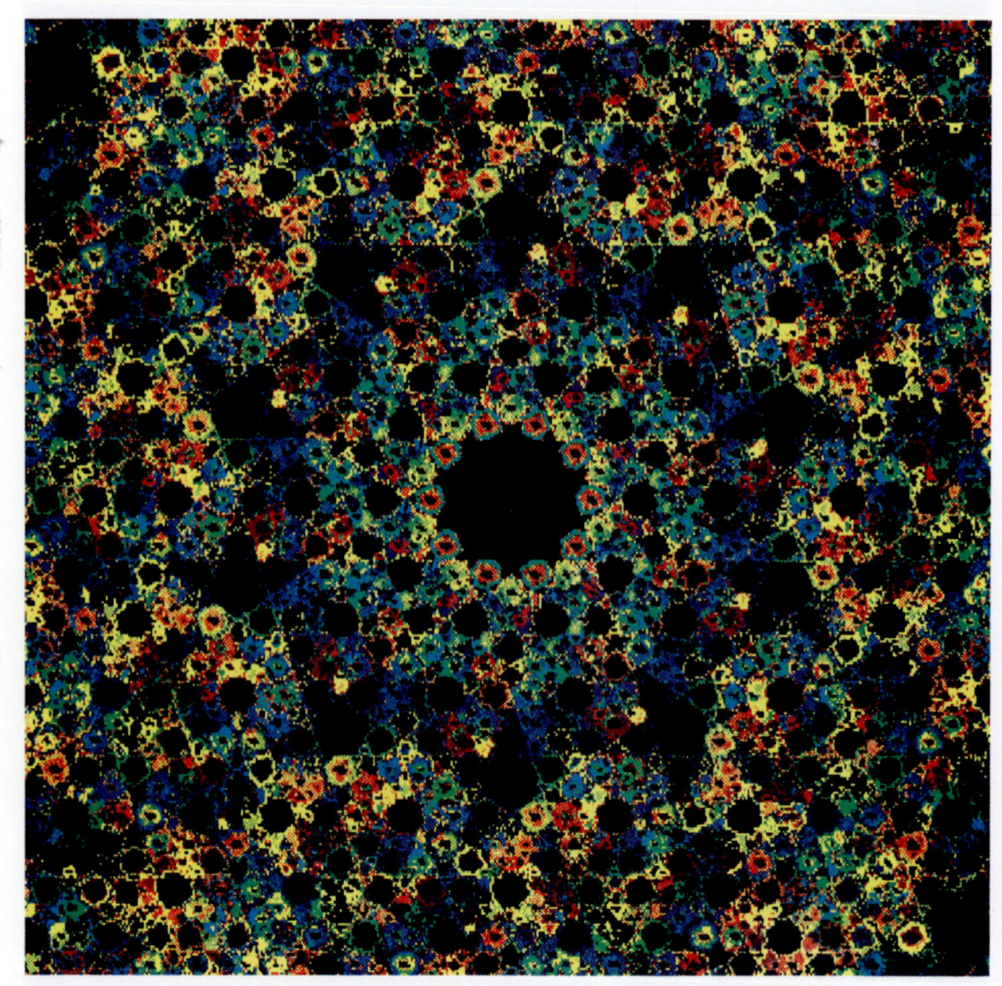

Plate C.3

Plate C.4

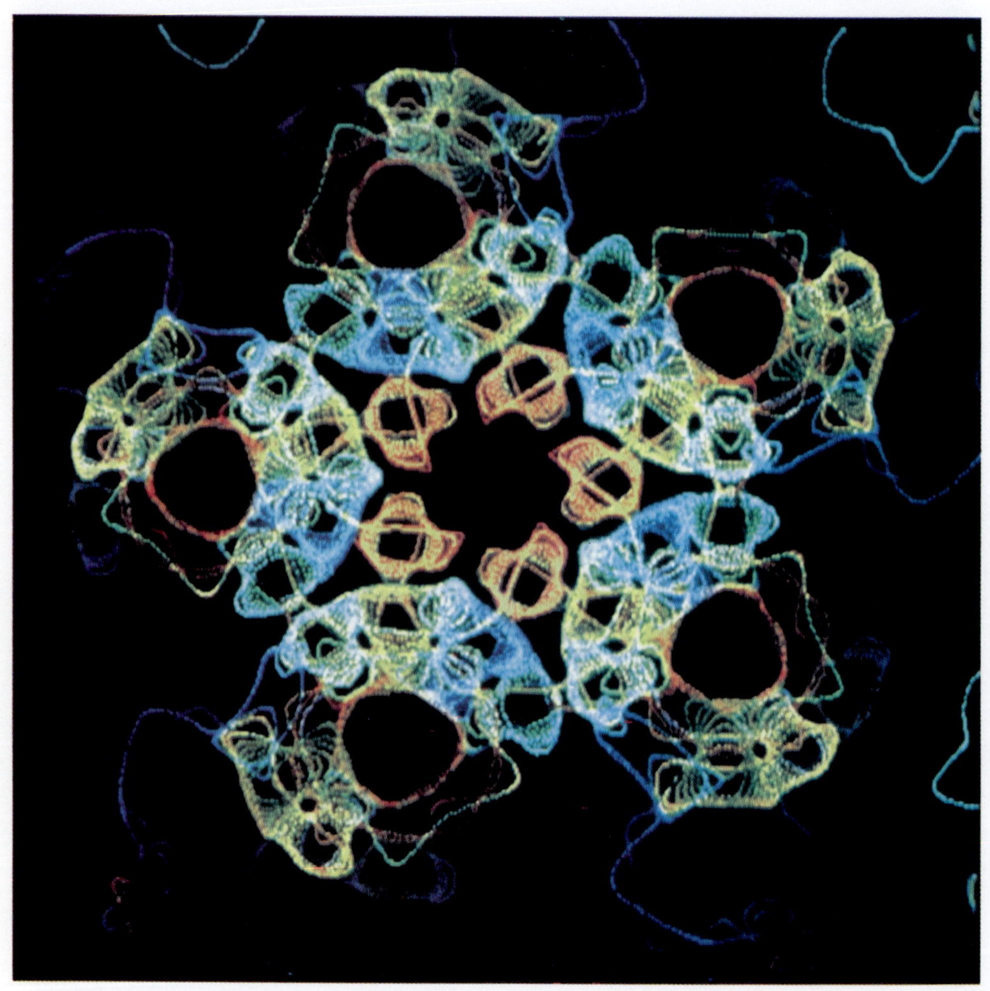

Plate C.5

Plate C.6

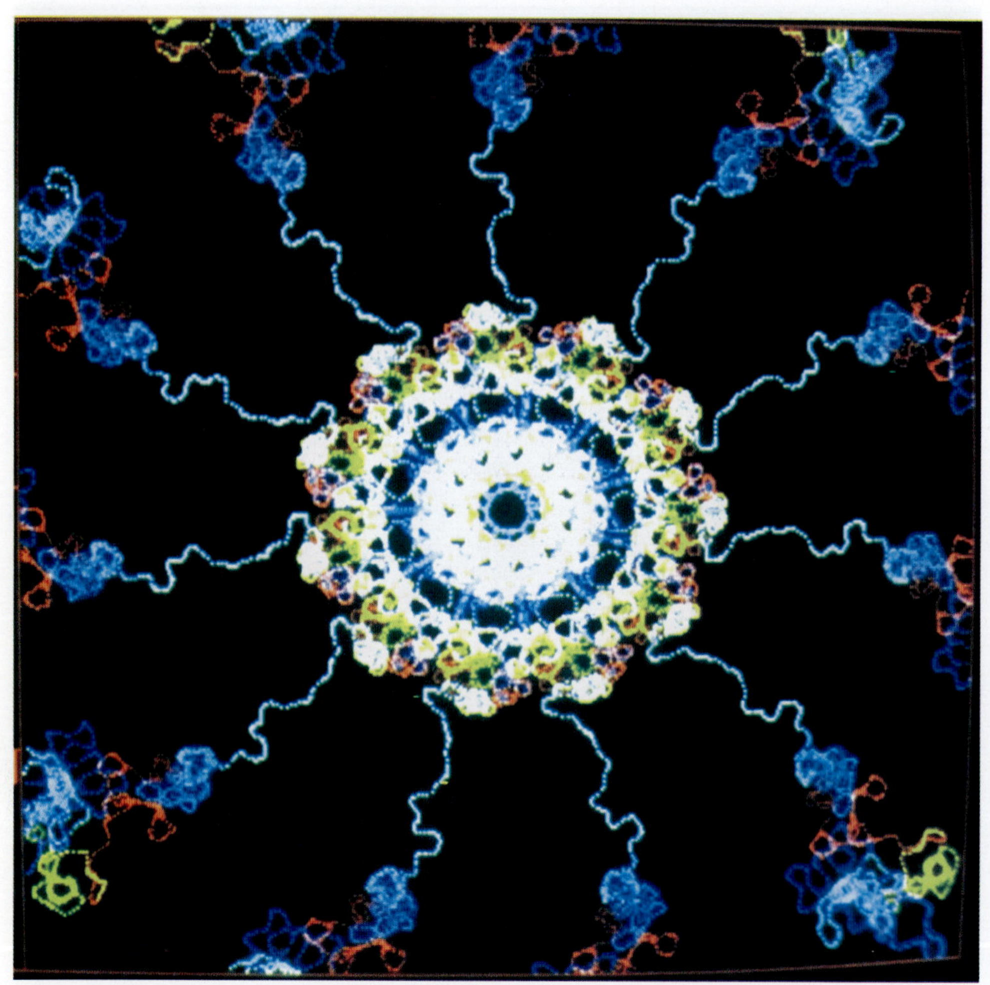

Plate C.7

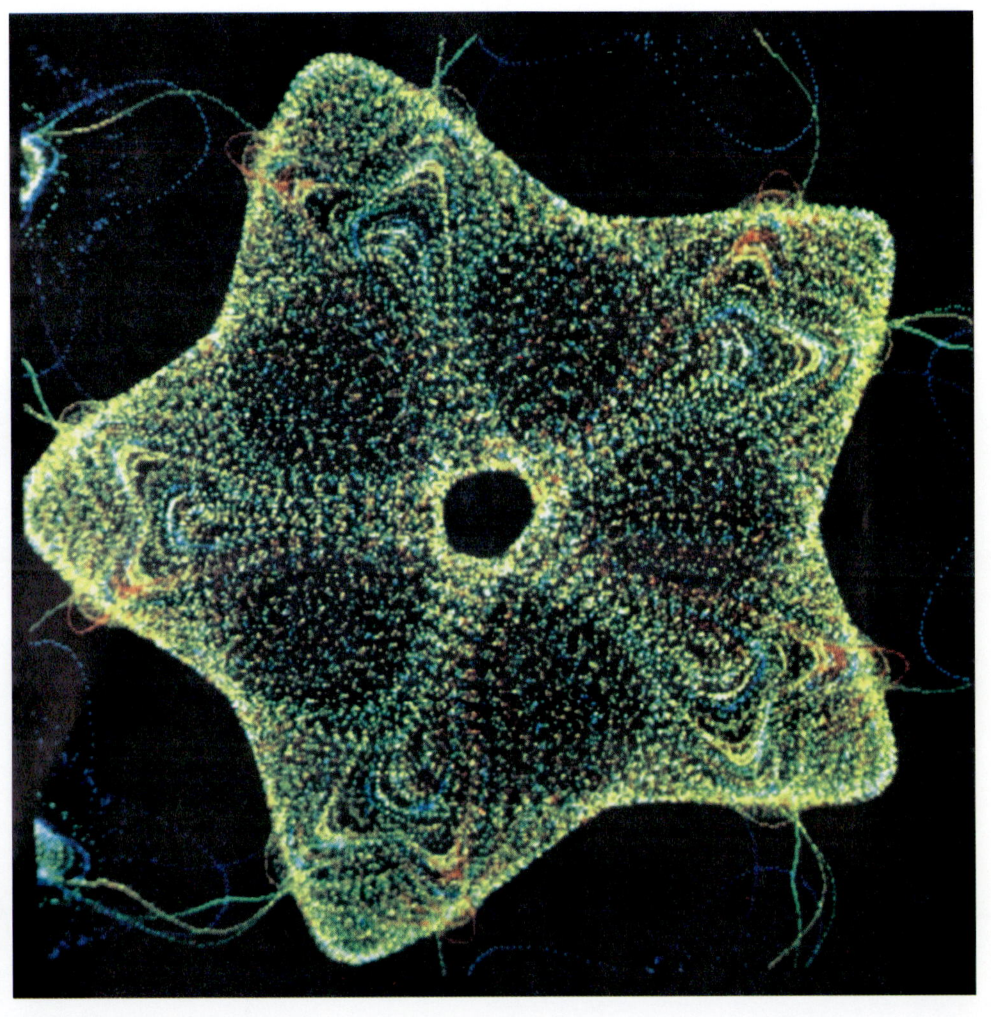

Plate C.8

# *Chapter 8*

# CHAOS AND SYMMETRY

## 8.1 Stochastic Webs

A system is close to an integrable one if its Hamiltonian can be presented as

$$H = H_0(I) + \epsilon V(I, \vartheta), \tag{8.1.1}$$

where $I = (I_1, \ldots, I_N)$ and $\vartheta = (\vartheta_1, \ldots, \vartheta_N)$ are the $N$-dimensional vectors of action and angle, respectively, and $\epsilon$ is a small parameter of perturbation. For $\epsilon = 0$, the system is described by its Hamiltonian $H_0$ and possesses $N$ first commuting integrals of motion (actions). The motion is performed along an $N$-dimensional torus defined by a specific value of $I$. In the set of tori, some of them are singular and correspond to separatrices or hypersurfaces with self-intersections. In two-dimensional phase space, separatrices are special trajectories that pass through the saddle point(s). The set of separatrices can form a connected net in the phase space (see Fig. 8.1.1) where a particle can move along. In fact, the likelihood of a particle moving along a separatrix is minimal since the time taken to reach a saddle point is infinite and it is not able to pass through the first saddle. Inside the meshes of the net, the motion is finite. This leads one to conclude that the general type of integrable motion is partitioned by a net of separatrices in an integrable case, in addition to the Liouville-Arnold theorem on integrability.[1]

[1]A rigorous formulation of the integrability conditions is very important. This can be found in [A 89]. A non-trivial example of the violation of the conditions formulated in [A 89] is found in [RB 81].

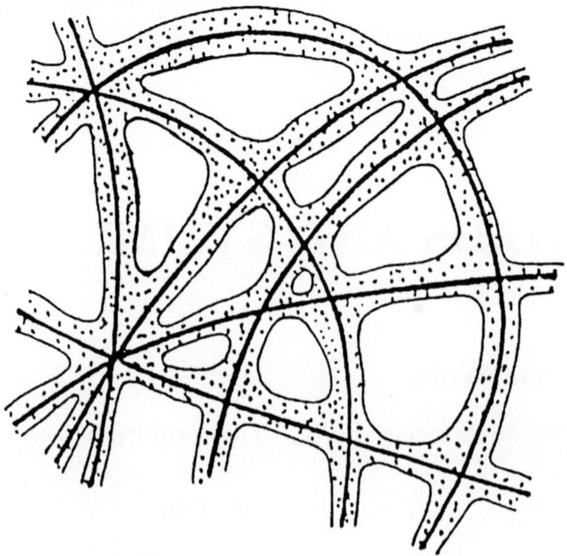

Fig. 8.1.1. Stochastic web formation.

The influence of a small perturbation in (8.1.1) does not change significantly the motion inside the meshes of the net. However, it causes a crucial replacement of the separatrices by stochastic layers. A connected net of channels of finite width along which an effective particle transport can be performed is therefore created. The set of channels is called a stochastic web (see Fig. 8.1.1).

Different kinds of finite or infinite stochastic webs are possible, including those with or without a symmetry. It is convenient to introduce a notion of the web's skeleton since the width of the web tends to zero if $\epsilon \to 0$. It is possible for the skeleton of the web not to coincide with the separatrices net for $\epsilon = 0$, and the web can only occur when there is perturbation. Such a case takes place for the periodically kicked oscillator problem (1.3.1),

$$H = \frac{1}{2}(\dot{x}^2 + \omega_0^2 x^2) - \frac{\omega_0}{T} K \cos x \sum_{m=-\infty}^{\infty} \delta\left(\frac{t}{T} - n\right), \qquad (8.1.2)$$

where the oscillator's mass is equal to one. The unperturbed part of the Hamiltonian is the free oscillator motion,

$$H_0 = \frac{1}{2}(\dot{x}^2 + \omega_0^2 x^2)\,, \tag{8.1.3}$$

either with a degenerate rotational symmetry or without any separatrices or saddle points. The perturbation has an amplitude $K$ and a translational symmetry along $x$: $x \to x + 2\pi m$ with integer $m$. As demonstrated in Section 1.3, when $K \neq 0$, stochastic webs occur if the resonant condition

$$\alpha = \omega_0 T = 2\pi/q \tag{8.1.4}$$

is valid, where $q$ is an integer. Although this is not the only case claiming the existence of stochastic webs, (8.1.4) is nevertheless adhered to for the sake of simplicity.

The general comments on the web problem are:

(i) The existence of stochastic webs is an important physical phenomenon since infinite particle transport is performed along the channels of the web.
(ii) The skeleton of the web tiles the phase plane (or volume) and imposes a dynamical origin on symmetry groups.

For example, the Arnold diffusion occurs for the number of degrees of freedom $N > 2$. It requires the non-degeneracy condition

$$\left|\frac{\partial^2 H_0}{\partial I_j \partial I_k}\right| \neq 0\,. \tag{8.1.5}$$

The diffusion rate is very low. Stochastic webs that occur in the model (8.1.2) correspond to the case when (8.1.5) is not valid and $N < 2$. The diffusion rate along the channels of the web is much faster than in the case of Arnold diffusion. Many other possibilities also exist for higher dimensions ($N > 2$). Some of them will be considered in this chapter.[2]

[2]The Arnold diffusion was introduced in [A 64]. Some proofs were published in [Ne 77]. The different physical applications of the Arnold diffusion can be found in [C 79] and [LiL 92].

## 8.2 Stochastic Web with Quasi-Crystalline Symmetry

The stochastic webs considered below occur when the condition of non-degeneracy (8.1.5) is violated. When the Hamiltonian (8.1.2) is rewritten using the more convenient variables (1.3.6) $u = \dot{x}/\omega_0$ and $v = -x$, $H$ is renormalised by a constant,

$$H = \frac{\alpha}{2}(u^2 + v^2) - K \cos v \sum_{n=-\infty}^{\infty} \delta(\tau - n)\,, \tag{8.2.1}$$

with equation of motion

$$\dot{u} = \frac{\partial H}{\partial v}\,, \quad \dot{v} = -\frac{\partial H}{\partial u} \tag{8.2.2}$$

where the dot denotes a derivative with respect to the dimensionless time $\tau = t/T$. The equations of motion (8.2.2) can be partially integrated (as it was done in Section 1.3) in order to replace them by the discrete map (1.3.5):

$$\hat{M}_\alpha: \quad \begin{array}{l} u_{n+1} = (u_n + K \sin v_n)\cos\alpha + v_n \sin\alpha \\ v_{n+1} = -(u_n + K \sin v_n)\sin\alpha + v_n \cos\alpha \end{array}\,. \tag{8.2.3}$$

Here, one considers the resonance case (8.1.4) which defines the web-map

$$\hat{M}_q: \quad \begin{array}{l} u_{n+1} = (u_n + K \sin v_n)\cos\dfrac{2\pi}{q} + v_n \sin\dfrac{2\pi}{q} \\ v_{n+1} = -(u_n + K \sin v_n)\sin\dfrac{2\pi}{q} + v_n \sin\dfrac{2\pi}{q} \end{array}\,. \tag{8.2.4}$$

The map $\hat{M}_\alpha$ has a central elliptic fix point (0,0). The eigenvalues, $\lambda_\alpha$, of the tangent matrix, $\hat{M}'_\alpha$, at this point satisfy the equation

$$\lambda_\alpha^2 - 2Sp\hat{M}'_\alpha \cdot \lambda_\alpha + 1 = 0\,, \tag{8.2.5}$$

where $Sp\hat{M}'_\alpha$ refers to the sum of the digaonal elements of $\hat{M}'_\alpha$. The solution of (8.2.5) gives rise to the condition

$$K > 2|\text{cotan}(\alpha/2)| \equiv K_\alpha^{(0)} \tag{8.2.6}$$

when the origin (0,0) becomes unstable and the elliptic point is transferred to the hyperbolic one.

In Section 1.3, a few cases of the stochastic web with a crystalline symmetry, that is, when $q = 3, 4, 6$ and belonging to the set

$$\{q_c\} \equiv \{1, 2, 3, 4, 6\}, \tag{8.2.7}$$

are demonstrated. The set $\{q_c\}$ corresponds only to cases when the rotational and translational symmetries coexist together on the plane. The cases where $q = 1, 2$ are trivial and their corresponding motion is therefore regular. Figures 1.3.1 and 1.3.2 showed stochastic webs with the crystal-type symmetry of orders 4, 3 and 6. Some of their properties are formulated as follows:

(i) Crystal-type stochastic webs exist for arbitrarily small $K$. It does not change the size of the meshes since $K \to 0$ (up to the higher order terms in $K$, and it disappears when $K = 0$). Proof of this statement is found in Section 8.4.

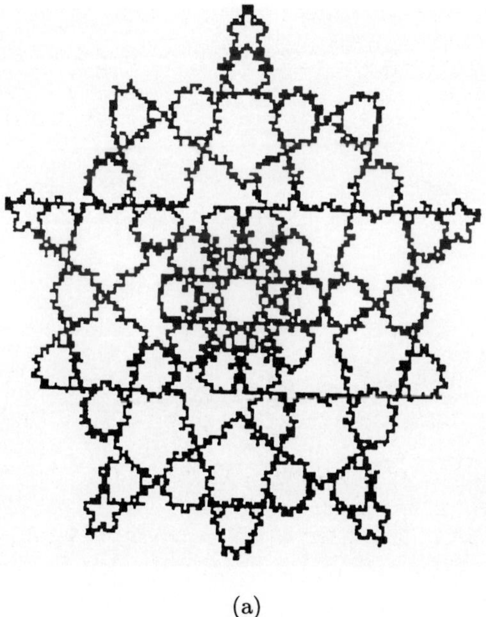

(a)

Fig. 8.2.1. Three different samples of the five-fold symmetric stochastic web.

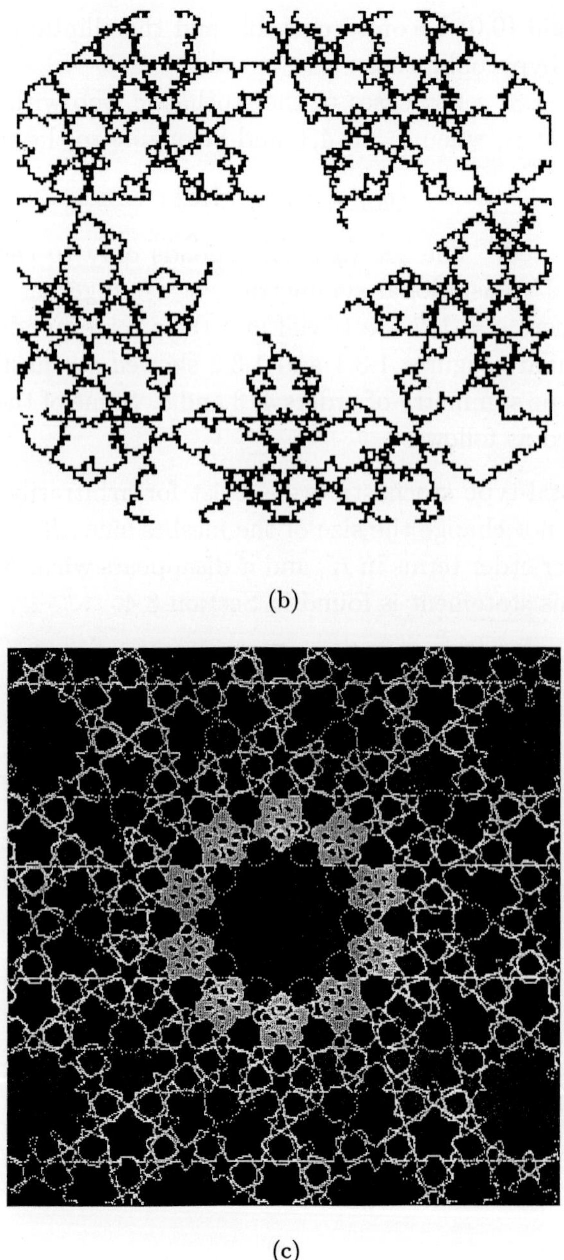

(b)

(c)

Fig. 8.2.1. (*Continued*)

(ii) Diffusion along the webs is proportional to $K^2$ unless an anomalous transport takes place. This statement is proven in Chapter 9.

A stochastic web will still exist for $q \not\equiv \{q_c\}$ if $1 > K > K_c$, where $K_c$ is a small critical value. Examples of the corresponding webs for $q = 5$ are shown in Fig. 8.2.1 and in Fig. 8.2.2 when $q = 7$ and 8. In all these figures, the webs are infinite and their finite sizes correspond to a finite computing time. All the webs have the quasi-crystal type symmetry which will be explained in the next section. It can be assumed that

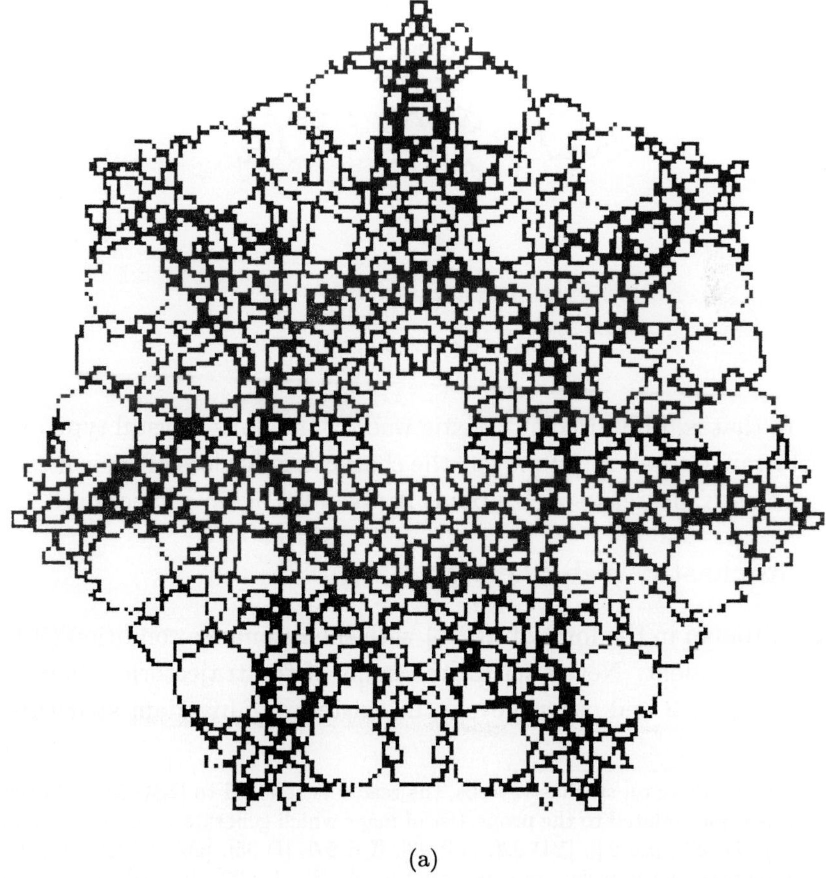

(a)

Fig. 8.2.2. Stochastic webs of the (a) seven-fold and (b) eight-fold symmetry.

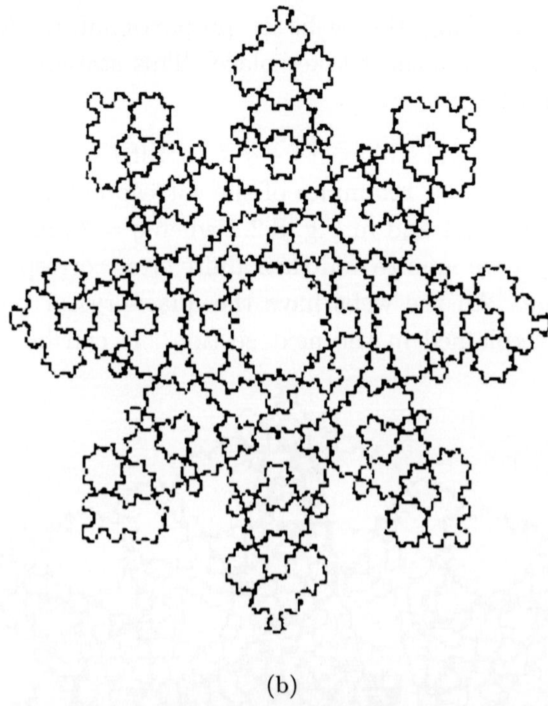

(b)

Fig. 8.2.2. (*Continued*)

$K_c = 0$, that is, unlimited stochastic webs of the quasi-crystal type exist for arbitrarily small $K$. However, the characteristic sizes of their meshes increase as $K$ decreases. This point is taken up again in Section 8.4.[3]

## 8.3 Stochastic Web Skeleton

A Hamiltonian in the form of (8.2.1) with the resonance condition (8.14) is time-dependent. Nevertheless, the maps of the trajectories shown in Figs. 8.2.1, 8.2.2 and others reveal the existence of invariant structures.

[3]For further reading on stochastic webs, the reader is referred to [ZSUC 91]. Several interesting topics related to the properties of maps which generate webs can be found in [Lo 91], [Lo 92], [Lo 94], [YP 92], [YP 93], [PR 97], [D 95], [DA 95] and [DK 96]. The symmetries of the webs are considered in [A 88], [Lo 92], [Lo 93], [La 93] and [LQ 94].

By applying an appropriate method of averaging, an effective Hamiltonian, $H_q(u, v)$, which describes the structures, is obtained and it should be time-independent and integrable.

When the polar co-ordinates $(\rho, \phi)$ are introduced,

$$u = \rho \cos \phi\,, \quad v = -\rho \sin \phi\,. \tag{8.3.1}$$

The generating function

$$F = (\phi - \alpha\tau)I\,, \quad I = \frac{1}{2}\rho^2 \tag{8.3.2}$$

corresponds to the transformation of a co-ordinate system rotating at a frequency of $\alpha$. In terms of the new variables, the new Hamiltonian becomes

$$\tilde{H} = H + \frac{\partial F}{\partial \tau} = -K \cos[\rho \cos(\phi - \alpha\tau)] \sum_{n=-\infty}^{\infty} \delta(\tau - n)\,, \tag{8.3.3}$$

where the first term of the Hamiltonian (8.2.1), which represents linear oscillations, disappears. This is a typical feature of the transformation into a rotating frame of reference.

A simple transformation of the sum of the $\delta$ functions in Eq. (8.3.3) is performed:

$$\sum_{n=-\infty}^{\infty} \delta(\tau - n) = \sum_{j=1}^{q} \sum_{m=-\infty}^{\infty} \delta[\tau - (mq + j)] \tag{8.3.4}$$

and the representation for the $\delta$-functions sum is used:

$$\sum_{m=-\infty}^{\infty} \delta(\tau - j - mq) = \frac{1}{q} \sum_{m=-\infty}^{\infty} \exp\left(2\pi i m \frac{\tau - j}{q}\right). \tag{8.3.5}$$

The substitution of (8.3.4) and (8.3.5) into (8.3.3) and a regrouping of terms yield

$$\tilde{H} = H_q + V_q$$

$$H_q = -\frac{1}{q} K \sum_{j=1}^{q} \cos \xi_j \tag{8.3.6}$$

$$V_q = -\frac{2}{q} K \sum_{j=1}^{q} \cos \xi_j \sum_{m=1}^{q} \cos \left( \frac{2\pi m}{q} (\tau - j) \right),$$

where

$$\xi_j = -\rho \sin \left( \phi + \frac{2\pi}{q} j \right) = v \cos \left( \frac{2\pi}{q} j \right) + u \sin \left( \frac{2\pi}{q} j \right). \tag{8.3.7}$$

Expression (8.3.7) can also be written in the compact form of

$$\xi_j = \boldsymbol{\rho} \cdot \mathbf{e}_j \tag{8.3.8}$$

using a unit vector definition

$$\mathbf{e}_j = [\cos(2\pi j/q), -\sin(2\pi j/q)] \tag{8.3.9}$$

and

$$\boldsymbol{\rho} = (v, u).$$

Equation (8.3.9) defines a regular star formed by $q$ unit vectors $\mathbf{e}_j$ and Eq. (8.3.8) defines the projection of the unit star onto a two-dimensional plane.

Expressions (8.3.6) and (8.3.8) define the stationary Hamiltonian

$$H_q \equiv H_q(u, v) = -\frac{1}{2} \Omega_q \sum_{j=1}^{q} \cos(\boldsymbol{\rho} \cdot \mathbf{e}_j), \quad \Omega_q \equiv \frac{2K}{q} \tag{8.3.10}$$

which is termed the skeleton Hamiltonian. Another part of $\tilde{H}$ is $V_q$, which defines the non-stationary part of the Hamiltonian $\tilde{H}$. The expression $H_q = H_q(u, v)$ defines a surface. The isolines of a constant energy,

$$E = -\frac{1}{2} \Omega_q \sum_{j=1}^{q} \cos(\boldsymbol{\rho} \cdot \mathbf{e}_j), \tag{8.3.11}$$

correspond to the invariant curves of the initial perturbed Hamiltonian (8.2.1) up to higher order terms. For example, if $q = 4$,

$$H_4 = -\Omega_4(\cos u + \cos v) \tag{8.3.12}$$

and the corresponding isolines are shown in Fig. 8.3.1. The values of $E > 0$ define sections of humps on the surface of $E = H_4(u, v)$, the values of $E = H_4(u, v) < 0$ define sections of wells, and the value $E = H_4(u, v) = E_s = 0$ defines an infinite net of separatrices that tile the plane with a four-fold symmetry. This net is the skeleton of the stochastic web for $q = 4$.

Fig. 8.3.1. The phase portrait for the four-fold symmetric skeleton Hamiltonian.

Similarly, one can obtain expressions when $q = 3$ and 6:

$$H_6 = 2H_3 = -\frac{1}{2}\Omega_6\left[\cos v + \cos\left(\frac{1}{2}v + \frac{\sqrt{3}}{2}u\right) + \cos\left(\frac{1}{2}v - \frac{\sqrt{3}}{2}u\right)\right]. \tag{8.3.13}$$

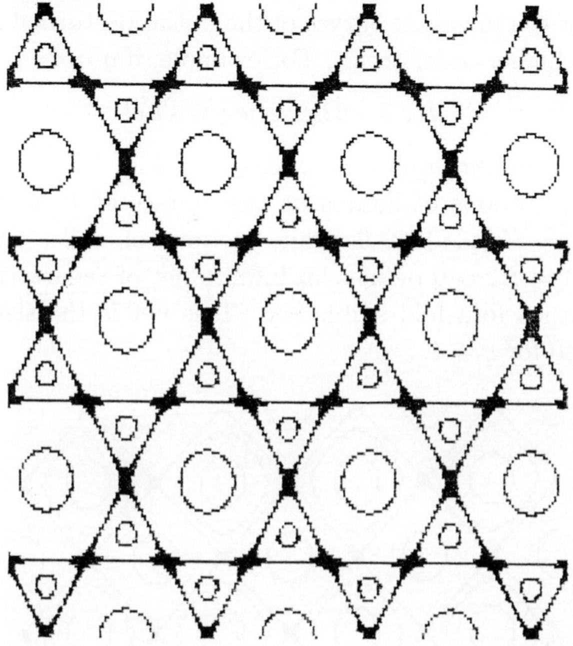

Fig. 8.3.2. The phase portrait for the 3(6)-fold symmetric skeleton Hamiltonian.

The structure of isolines is shown in Fig. 8.3.2. The net of separatrices can be obtained from the equation

$$E = E_s = \frac{1}{2}\Omega_6 = H_6(v_s, u_s) \tag{8.3.14}$$

and all solutions are

$$\begin{aligned} &v_s = \pi(2n_1 + 1)\,, \quad v_s = \sqrt{3}u_s + 2\pi(2n_2 + 1)\,, \\ &v_s = -\sqrt{3}u_s + 2\pi(2n_3 + 1)\,, \quad (n_{1,2,3} = 0, \pm 1, \ldots)\,. \end{aligned} \tag{8.3.15}$$

The net of separatrices at the skeleton of the stochastic web forms the so-called kagome lattice with hexagonal symmetry.

The general features of the skeletons for the crystalline symmetry ($q = 3, 4, 6$) are:

(i-c) All saddle points of the Hamiltonian $H_q(u, v)$ have the same value of energy $E_s$.

(ii-c) All separatrices are isolines of the same energy surface $H_q(u_s, v_s) = E_s$.

(iii-c) Skeletons are unbounded and tile the plane with the corresponding symmetry square or hexagon.

In the case of $q \not\equiv \{q_s\}$, the saddle points of the Hamiltonian $H_q(u, v)$ do not belong to some unique value of energy anymore. They are distributed in a complicated manner although some values of energy do exist

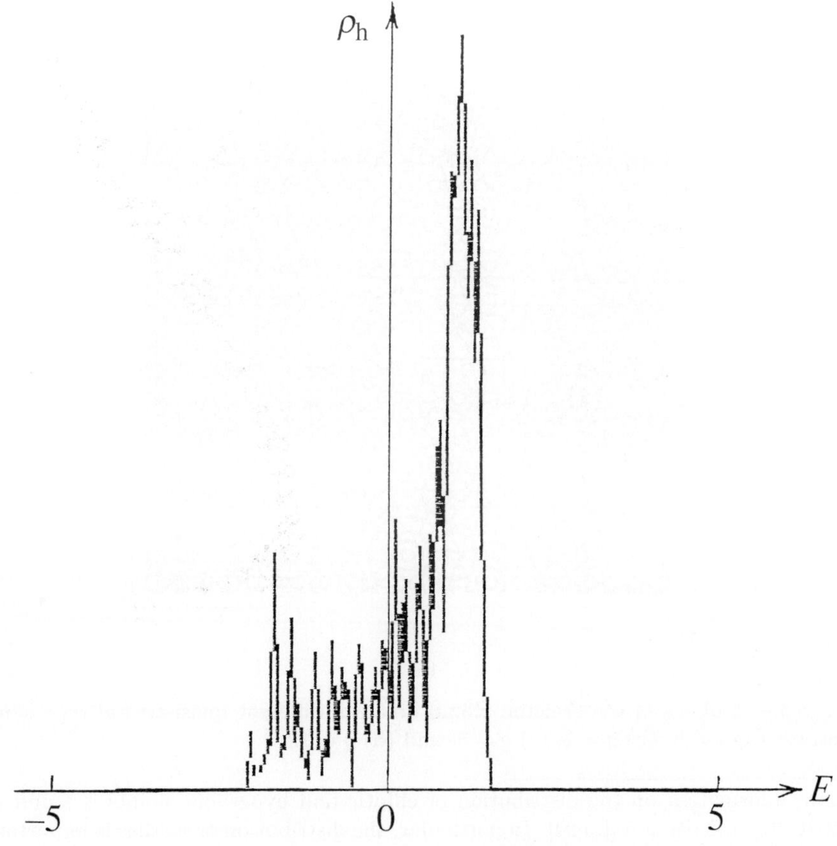


Fig. 8.3.3. Distribution of saddles for $H_5$.

where a higher concentration of saddle points is found. For example, when $q = 5$, the maximum number of saddle points corresponds to $E_m \approx -\frac{1}{2}\Omega_5$ (see Fig. 8.3.3). In fact, not much is known about the distribution of saddles and elliptic points for the values of $q \not\equiv \{q_c\}$.[4]

Consider, for example, the isolines for the values of energy $E$ which belong to a layer $E \in (E_m - \Delta E, E_m + \Delta E)$. They form a "thick" line structure. The corresponding pattern is shown in Fig. 8.3.4(a) and is related to the so-called quasi-crystal symmetry of the fifth order. The Hamiltonian $H_q$ in (8.3.10) is a general expression which generates quasi-crystal symmetry of isolines. The examples of isolines for $q = 7$, 8 and 12 are shown in Figs. 8.3.4(b), (c) and (d). They are obtained in the same

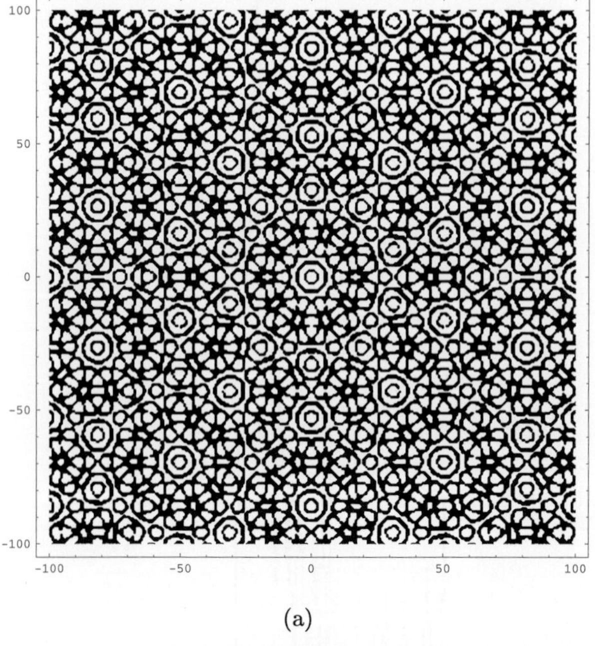


(a)

Fig. 8.3.4. Isolines of the skeleton Hamiltonian of different quasi-crystal type symmetries: (a) $q = 5$; (b) $q = 7$; (c) $q = 8$; and (d) $q = 12$.

[4] More information on the distribution of elliptic and hyperbolic points is found in [ZSUC 91], [Lo 91] and [Lo 94]. In particular, the distribution of saddles is important to the problem of percolation in the quasi-symmetric potential. For a discussion of the percolation problem along the webs, see [CZ 91] and [Is 92].

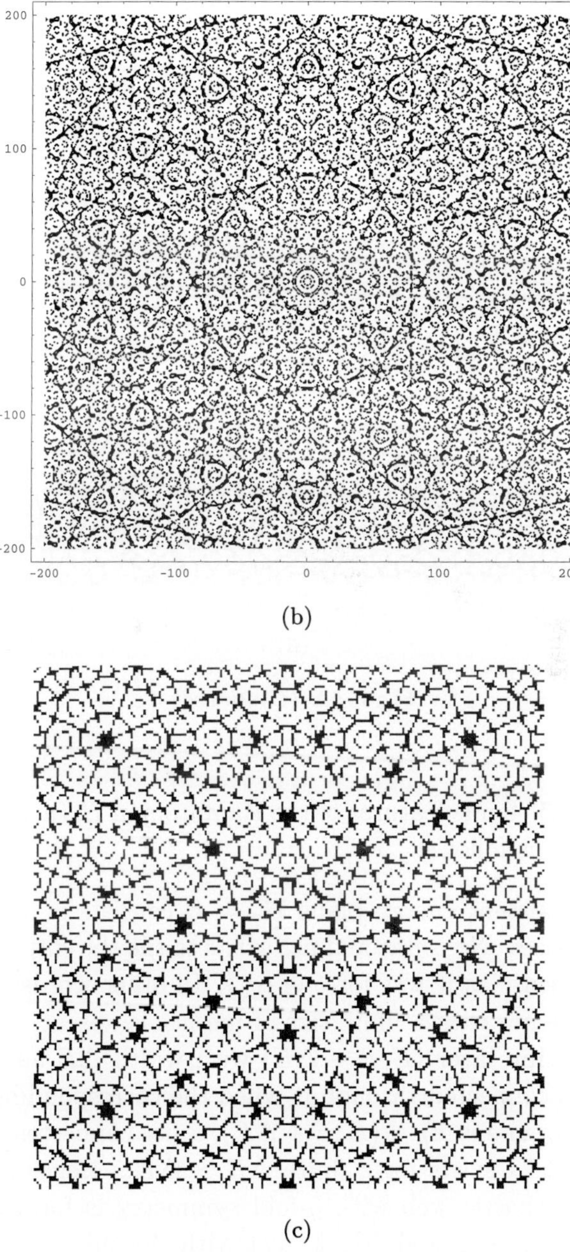


(b)

(c)

Fig. 8.3.4. (*Continued*)

(d)

Fig. 8.3.4. (*Continued*)

way as for $q = 5$. Some features of the Hamiltonian $H_q(u, v)$ for $q \notin \{q_c\}$ are formulated as follows:

(i-qc) Any constant energy plane $H_q = E$ consists of separatrices of finite sizes. No connected net exists to tile the entire plane in the same sense as it was for $q \in \{q_c\}$. In other words, a connected infinite net with a finite size of meshes is absent.

(ii-qc) For a thin layer $(E, E + \Delta E)$ with width $\Delta E$ and two saddle points connected through it when the corresponding energy for these points is $E_h \in (E, E + \Delta E)$, a net connected through the layer $\Delta E_c$ exists for a finite $\Delta E > \Delta E_c$. Hence a skeleton of the stochastic web with $q$-fold symmetry is formed. In other words, a connected "thick" net with the thickness $\Delta E_c$ exists. An example of "thick isolines" for $q = 5$ is shown in Fig. 8.3.5.

The connected net is easily visualised and it forms the so-called Ammann lattice.[5,6]

(iii-qc) It is assumed that the property (ii-qc) can be strengthened such that $\Delta E_c = 0$ and any finite thickness, $\Delta E > 0$, of the energy layer induces an infinitely connected net with a finite size for its meshes. The larger the meshes, the smaller the $\Delta E$ becomes.

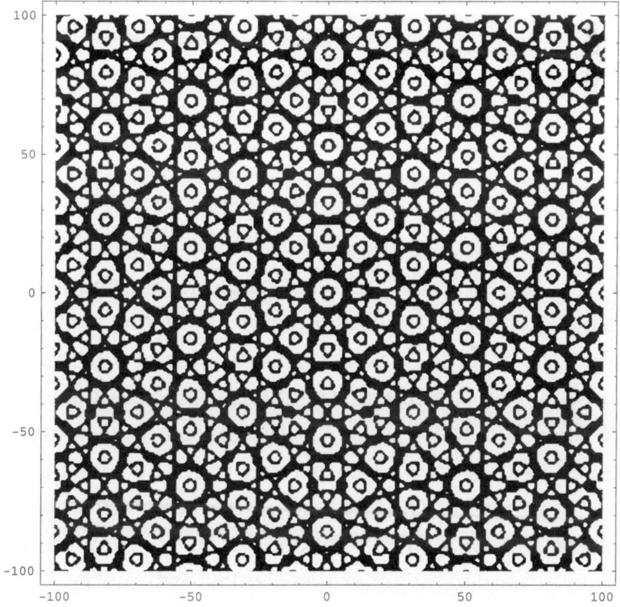


Fig. 8.3.5. "Thick isolines" for $q = 5$ in the energy layer $E \in (0, 2)$.

---

[5]The Ammann Lattice is a one-dimensional analog in the quasi-crystal with a five-fold symmetry. It is simply a one-dimensional "grid" of points with co-ordinates

$$x_n = n + \beta_1 + \frac{1}{\tau_0}\left[\frac{n}{\tau_0} + \beta_2\right],$$

where $n$ is an integer, $\beta_1$ and $\beta_2$ are any constants, $[a]$ is the integer part of $a$, and $\tau_0$ is the golden mean:

$$\tau_0 = (1 + \sqrt{5})/2 = 2\cos(\pi/5).$$

Information on the Ammann lattice is found in [GS 87].

[6]More information on the symmetries is found in [ZSUC 91], [La 93] and [LQ 94].

(iv-qc) The Hamiltonian $H_q$ has $q$-fold rotational symmetry for even $q$ and $2q$-fold rotational symmetry for odd $q$. The symmetry of $H_q$ does not coincide with the symmetry of the original Hamiltonian $H$ because of the averaging procedure. Up to the corresponding correction, the symmetries of $H$ and $H_q$ are similar for small $K$, and the symmetry properties of $H_q$ can be used to describe the real trajectories of the original Hamiltonian $H(x, \dot{x}, t)$.

(v-qc) The Hamiltonian $H_q$ defines an integrable system but there is no description of its solutions which follows from the equation [compare to (8.2.2)]:

$$\dot{u} = \frac{\partial H_q}{\partial v}, \quad \dot{v} = -\frac{\partial H_q}{\partial u}. \tag{8.3.16}$$

From all the properties described above, one can deduce the existence of the dynamical generator(s) of symmetry. It means that a Hamiltonian such as $H(u, v, t)$ in (8.2.1) is associated with a specific composition of orbits and singular points in the phase space (phase plane) which partitions or tiles the space with approximate $q$-fold symmetry (better known as quasi-symmetry). The accuracy of the quasi-symmetry can be enhanced by applying the averaging procedure to obtain a skeleton Hamiltonian, $H_q$, in (2.3.10). The smaller the parameter $K$ in (8.2.1), the closer to symmetry the generalised quasi-symmetry becomes.[7]

## 8.4 Symmetries and Their Dynamical Transforms

This section demonstrates how one can obtain some non-trivial symmetries from dynamics. It also shows how symmetries can be modified by using methods of dynamics. In considering how a five-fold symmetry can arise in dynamics, one resorts to using Hamiltonian (8.2.1) for $q = 5$, or its equivalent, and the map $\hat{M}_{q=5}$ from (8.2.4) as an example. This constitutes the first step. In the second step, the averaged Hamiltonian

[7]The problem of dynamical generation of symmetries was formulated by A. Weyl [We 52]. The web-map can be considered a solution to the problem of crystal and quasi-crystal symmetries in a two-dimensional case.

$\hat{H}_{q=5}$ in (8.3.10) is obtained. Next, the maximum saddle point distribution, as a function of energy and the corresponding value $E_{\text{max}}$, is derived. Fourth, the "thick isolines" portrait which belongs to the finite energy layer, $(E_{\text{max}} - \Delta E/2, E_{\text{max}} + \Delta E/2)$, with an appropriate small $\Delta E$ is found. Fifth, using the thick isolines portrait (see Fig. 8.3.4(a)) as a stencil and an algorithm to connect the different points on the stencil, a tiling is created. As an example, Fig. 8.4.1 demonstrates the famous Penrose tiling by connecting the centres of small five-stars with those of the small pentagons. A deliberate attempt was made to show a slightly asymmetrical tiling by using two kinds of rhombuses to illustrates different possibilities. A more sophisticated seven-fold symmetric tiling

Fig. 8.4.1. Formation of the Penrose tiling using Fig. 8.3.5 as a stencil.

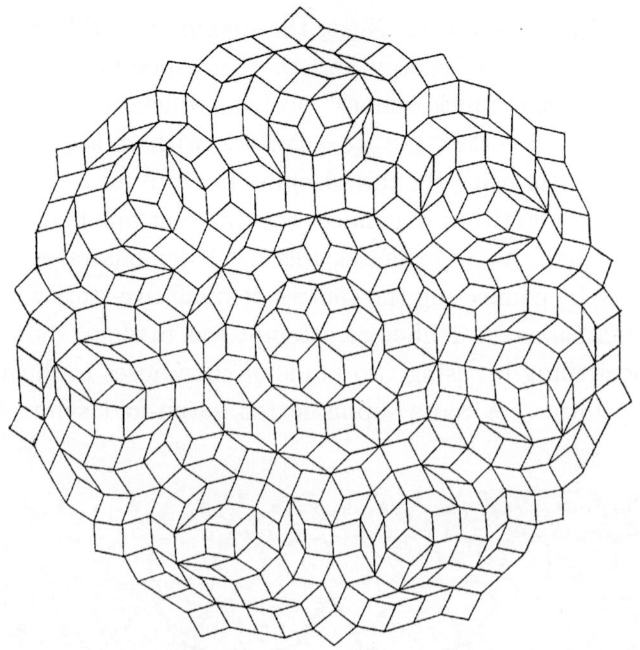

Fig. 8.4.2. Seven-fold symmetric tiling using three types of rhombuses.

with three kinds of rhombuses is shown in Fig. 8.4.2. Figure 8.3.4(b) has been used as a seven-fold symmetric stencil to produce the tiling in Fig. 8.4.2.[8]

The above comments suggest a new and unique opportunity of investigating a complex symmetry such as the quasi-crystalline symmetry by using methods of Hamiltonian dynamics. To demonstrate the relationship between symmetry breaking and bifurcation in a dynamic system, for $q = 3$, the invariant curves of the map $\hat{M}_3$ form a tiling of the hexagonal type if $K < K_\alpha^{(0)}$ with $\alpha = 2\pi/3$, where $K_\alpha^{(0)}$ is defined in (8.2.6). The tiling is shown in Fig. 8.4.3(a). From the same map $\hat{M}_3$, one can obtain a tiling of the brickwall type (Fig. 8.4.3(b)) if $K > K_\alpha^{(0)}$.

[8]For more information on the Penrose and other tilings, see [P 74], [Se 95] and [ZSUC 91].

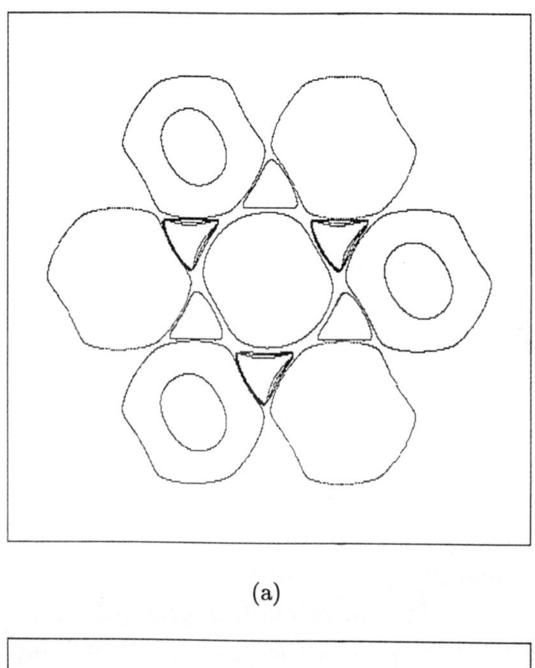

(a)

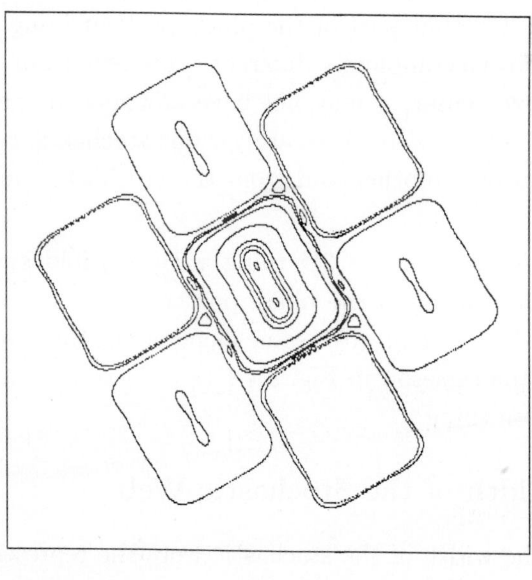

(b)

Fig. 8.4.3. Invariant curves for $q = 3$ form hexagonal tiling when (a) $K = 0.5$ and (b) brickwall tiling when $K = 1.3$.

The quantity $K > K_\alpha^{(0)}$ corresponds to the value of $K$ where the elliptic points at the centre of the hexagons lose their stability. For $q = 3$ or $\alpha = 2\pi/3$, $K_\alpha^{(0)} = 2\sqrt{3} = 1.1547\ldots$. If certain "fine" details of the figure are ignored, the hexagons are modified to become rectangles and there is a corresponding change in the symmetry of the stochastic web, which occupies very narrow regions between the rectangles, if $K > K_\alpha^{(0)}$. An example of a web with the brickwall symmetry is shown in Fig. 1.3.2(b).

The examples described in this section enable one to formulate the following statement: the web-map $\hat{M}_q$ is considered a dynamic generator for tiling a plane which has either a symmetry of the arbitrary order $q$ of the crystalline type when $q \in \{q_c\}$, or of the quasi-crystalline type when $q \notin \{q_c\}$. One of the tiling methods is related to obtaining a skeleton prior to performing an algorithm for the tiling (this method was described above). Another method uses the phase portrait of the system directly and it is simplified by removing some of the details or smoothing over some special orbits (for example, the stochastic web). It should be noted that part of the phase portrait lying inside a cell of the web may have a completely different symmetry from that of the web itself and, in particular, it may not have any special symmetry. Inside the central cell of the web, there are various stochastic layers which are separated from one another and from the stochastic web by invariant curves.

One can also say that a 2D tiling with a $q$-fold symmetry is the result of the decoration of the phase portrait. Decoration refers to the rule used to select and/or alter some elements of the phase portrait. A decoration of the skeleton in Fig. 8.4.1 ($q = 5$) results in its conversion into the Penrose tiling.

## 8.5 The Width of the Stochastic Web

To calculate the width of the stochastic web, the representation (8.3.6) for the Hamiltonian $\tilde{H}$ in the rotating frame of reference is used. The case of $q = 4$ provides a good illustration of the method described in Section 2.4.

Using Eq. (8.3.6), one obtains

$$\begin{aligned} \tilde{H} &\approx H_4 + V_4 \\ H_4 &= -\Omega_4(\cos u + \cos v) \\ V_4 &= -2\Omega_4(\cos v - \cos u)\cos \pi\tau \end{aligned} \tag{8.5.1}$$

by neglecting the other terms in $V_4$ which are proportional to $\cos(m\pi\tau)$ with $m > 1$. The small influence they exert will become apparent at the end of the calculations. Due to the degeneracy of the unperturbed Hamiltonian in (8.2.1), no small parameter is found in the perturbation $V_4$ in (8.5.1) and the influence of the perturbation is small via a different mechanism (as it was, for example, in Section 2.7).

For the unperturbed Hamiltonian $H_4$ defined in (8.3.11), the energy value on the separatrices is $H_4 = 0$ and the corresponding separatrices are defined by the equations

$$v_s = \pm(u_s + \pi) + 2\pi m\,, \quad (m = 0, \pm 1, \ldots)\,. \tag{8.5.2}$$

From the equation of motion (8.3.16), one derives

$$\dot{u} = \Omega_4 \sin v\,, \quad \dot{v} = -\Omega_4 \sin u\,. \tag{8.5.3}$$

On the separatrices

$$\dot{u}_s = \mp\Omega_4 \sin u_s\,, \quad \dot{v}_s = \pm\Omega_4 \sin v_s \tag{8.5.4}$$

or

$$\begin{aligned} \tan(u_s/2) &= e^{\mp\Omega_4(\tau - \tau_n)} \\ \tan(v_s/2) &= e^{\mp\Omega_4(\tau - \tau_n)}\,, \end{aligned} \tag{8.5.5}$$

where $\tau_n$ is the point of reference in time, (8.5.5) can give rise to

$$\sin u_s = -\sin v_s = -1/\cosh[\Omega_4(\tau - \tau_n)]\,. \tag{8.5.6}$$

Adopting the same scheme used in Sections 2.3 and 2.4, Eq. (8.5.1) yields

$$\dot{H}_4 = 4\Omega_4^2 \sin u \cdot \sin v \cdot \cos \pi\tau\,. \tag{8.5.7}$$

Replacing $u, v$ on the right-hand side with $u_s, v_s$ and substituting (8.5.6) in (8.5.7) as well as performing the integration, one obtains a change in the energy due to the perturbation in the particle motion in the vicinity for the separatrix:

$$\Delta H_4 = -4\Omega_4^2 \int_{\tau_n-\infty}^{\tau_n+\infty} d\tau \frac{\cos \pi\tau}{\cosh^2 \Omega_4(\tau - \tau_n)} = -4\pi^2 \frac{\cos \pi\tau_n}{\sinh(\pi^2/2\Omega_4)} . \tag{8.5.8}$$

Since $\Omega_4 = K/2 \ll 1$, this expression takes the form of

$$\Delta H_4 = -8\pi^2 \cos \pi\tau_n \exp(-\pi^2/K) . \tag{8.5.9}$$

The period of oscillations in the vicinity of the separatrix is

$$T(H_4) = \frac{4}{\Omega_4} \ln(8\Omega_4/|H_4|) = \frac{8}{K} \ln \frac{4K}{|H_4|} . \tag{8.5.10}$$

The time needed to pass near the separatrix is equivalent to one-quarter of the period, that is,

$$\tau_{n+1} - \tau_n \approx T(H_4)/4 = \frac{1}{\Omega_4} \ln(8\Omega_4/|H_4|) = \frac{2}{K} \ln \frac{4K}{|H_4|} . \tag{8.5.11}$$

Setting $H_4$ in the formula as being equivalent to $H_4^{(n+1)}$, and taking the expression (8.5.8) into account, a separatrix map in the form of

$$\begin{aligned} H_4^{(n+1)} &= H_4^{(n)} - 8\pi^2 \exp(-\pi^2)/K) \cos \pi\tau_n \\ \tau_{n+1} &= \tau_n + \frac{2}{K} \ln(4K/|H_4^{(n+1)}|) \end{aligned} \tag{8.5.12}$$

is derived. This procedure was described in Sections 2.3 and 2.4. Using the same procedure, the boundary of the stochastic layer from the condition

$$\max \left| \frac{\delta\tau_{n+1}}{\delta\tau_n} - 1 \right| > 1 \tag{8.5.13}$$

is found. Hence for the border $H_s$,

$$H_s = \frac{16\pi^3}{K} \exp(-\pi^2/K) \tag{8.5.14}$$

is obtained.[9]

The quantity $2H_s$ defines the width of the layer. Since all separatrices are connected, hence forming a single network, the quantity $2H_s$ is also the width of the stochastic web. It is exponentially small since $K \ll 1$. The neglected terms arising from $V_4$ in (8.5.1) give the corrections for the order $O[\exp(-2\pi^2/K)]$, which are too small.

The same considerations can be applied in the case of an arbitrary value of $q$. The width of a stochastic web can be written as

$$H_s \sim \exp(-\text{const}/K)\,, \tag{8.5.15}$$

where the value of const. increases somewhat with the growth of $q$. However, for all $q \notin \{q_c\}$, there is a peculiarity which needs to be addressed.

It was mentioned in Section 8.3 that all hyperbolic points are found on the same plane of the energy level $E$ for $q \in \{q_c\}$. In the case of $q \notin \{q_c\}$, this property is absent. Saddles and separatrices are distributed among different planes of the energy level and there is no single infinite net of separatrices. An example of isolines for $H_5$ is presented in Fig. 8.5.1 for $q = 5$. The figure shows saddles forming a pattern resembling a family of parallel straight lines (a). However, in Fig. 8.5.1(b), it can be seen that near the value $E_m = -\Omega_5/2$ (see Fig. 8.3.3), the separatrix loops approach each other very closely. Nevertheless, the intersections of the loops are located on different planes of constant energy. The gaps between some separatrices can be so small that even a small perturbation can create stochastic layers to cover these gaps. This is how a single stochastic web is formed. In the case of quasi-symmetry, the mechanism for the web's formation is quite different from that for crystal symmetry since there is no single separatrix for the web-skeleton Hamiltonian $H_q$.

[9]The results described were obtained in [ASZ 91].

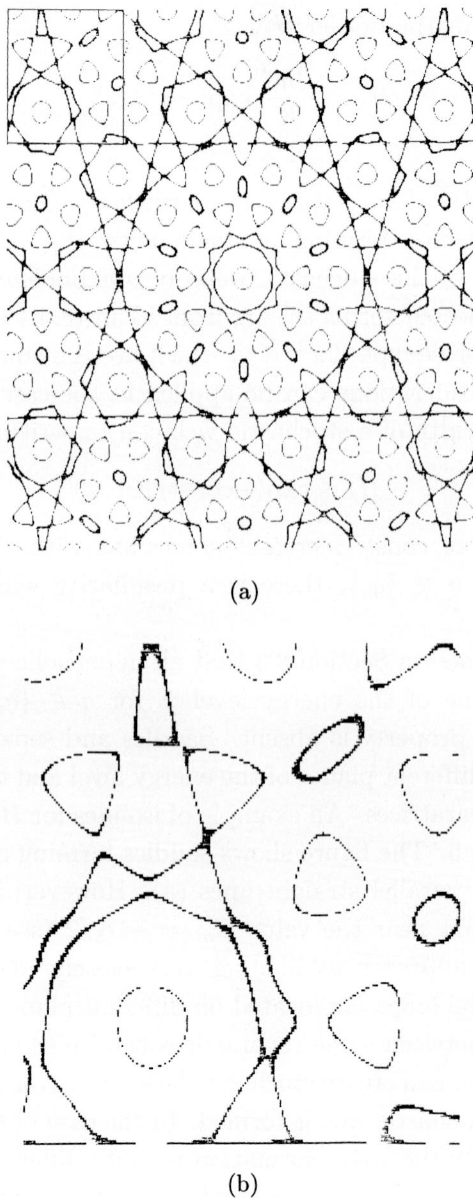


Fig. 8.5.1. Example of isolines for $H_5$: (a) the "thick lines" belong to the range of energy $E \in (0.8, 1.2)$; (b) an enlarged image of isolines from the marked square in (a) with $E \in (0.9, 1.1)$.

## Conclusions

1. The topological properties of the phase plane are defined by a set of singular points and curves. These properties can be imposed by an unperturbed Hamiltonian and a perturbation. The symmetry of the Poincaré map is the result of a complex interplay between the symmetries of the unperturbed Hamiltonian and the perturbation.

2. Thin filaments (channels) of chaotic dynamics are called stochastic web. They can form a net using symmetry or quasi-symmetry. The web-map discussed in Chapter 1 can tile the phase space with a symmetry of the quasi-crystal type.

3. Tiling can be characterised by a web skeleton. The web-map is a generator of the $q$-fold symmetric tiling of an arbitrary order $q$. The effective Hamiltonian for obtaining the web skeleton is

$$H_q(u,v) = \sum_{j=1}^{q} \cos(\mathbf{r} \cdot \mathbf{e}_j)$$

   with

$$\mathbf{r} = (u,v)\,, \quad \mathbf{e}_j = \left(\cos\frac{2\pi}{q}j, \sin\frac{2\pi}{q}j\right).$$

   The periodic tiling corresponds to $q = 1, 2, 3, 4, 6$. The rest are aperiodic and are equivalent to the two-dimensional quasi-crystal structure.

4. It is significant that the connected net with $q$-fold symmetry appears as a structure consisting of finite width channels in the phase plane. The width of the channels is exponentially small.

# Chapter 9

# MORE DEGREES OF FREEDOM

## 9.1 General Remarks

The previous chapters have considered systems with $N = 1\ 1/2$ degrees of freedom, that is, systems with one degree of freedom perturbed by a periodic in time force since the time is considered an additional "half" degree of freedom (phase). When one more degree of freedom is added, $N = 21/2$ is obtained. It follows from the KAM theory that there should be a qualitative change in the dynamics if the condition of non-degeneracy

$$J \equiv \left| \frac{\partial^2 H_0}{\partial I_j \partial I_k} \right| \neq 0 \tag{9.1.1}$$

is valid. This statement is related to the Arnold diffusion which exists if $N > 2$. Nevertheless, typical of many physical systems, the opposite condition $J = 0$, or $J$ is very close to zero, is present. This possibility has been discussed in detail in Sections 4.1 and 4.4. The condition (9.1.1) is violated in all systems similar to the linear oscillator without perturbation or in systems for which the nonlinear frequency

$$\omega_k = \partial H_0 / \partial I_k \tag{9.1.2}$$

is a non-monotonic function. For systems with the lifted (9.1.1) condition, or with $J$ close to zero described in Section 4.4, there are significant new properties of diffusion along the thin channels of the phase space where the dynamics are chaotic, that is, along the stochastic webs. The difference is characterised by two features. For $J = 0$, an infinite diffusion exists even for $N = 1\ 1/2$ while the Arnold diffusion occurs only for $N > 2$. The diffusion rate is also much higher for $N > 2$ than in the case of the Arnold diffusion if, for at least one degree of freedom, there is degeneracy or the corresponding frequency $\omega_j(I)$ is non-monotonic. This feature will be elaborated further when common physical models are introduced.

## 9.2 Four-Dimensional Map for the Motion in Magnetic Field

In this section, a four-dimensional map corresponding to a coupled web-map and a standard map is introduced. The map is considered a typical problem of dynamics in 4D phase space since both the web-map and the standard map correspond to two complimentary cases in 2D phase space.

For particle dynamics in a constant magnetic field, $B_0$, oriented along the $z$-axis and in a wave packet, $\mathbf{E}(\mathbf{r}, t)$, that propagates obliquely to $\mathbf{B}_0$, the corresponding equation of motion is

$$\ddot{\mathbf{r}} = \frac{e}{m_0}\mathbf{E}(\mathbf{r}, t) + \frac{e}{m_0 c}[\dot{\mathbf{r}}, \mathbf{B}_0]\,, \tag{9.2.1}$$

where $\mathbf{B}_0$ lies along the $z$-axis. The electric field, $\mathbf{E}(\mathbf{r}, t)$, is found on the $xz$-plane and is chosen as

$$\begin{aligned} \mathbf{E}(\mathbf{r}, t) &= -\mathbf{E}_0 \sum_{n=-\infty}^{\infty} \sin(\mathbf{k}r - n\Delta\omega t) \\ &= -\mathbf{E}_0 T \sin(k_x x + k_z z) \sum_{n=-\infty}^{\infty} \delta(t - nT)\,. \end{aligned} \tag{9.2.2}$$

This means that the wave packet has spectral uniformity with sufficiently large spectral width. The time interval, $T = 2\pi/\Delta\omega$, is determined by the frequency interval of the harmonics in the packet. For the longitudinal electric field, the wave vector $\mathbf{k}$, as well as the amplitude vector $\mathbf{E}_0$, has only two components, $k_x$ and $k_z$. They are related by

$$k_z/k_x = E_{0z}/E_{0x} = \beta = \text{const.} \tag{9.2.3}$$

Rewriting Eq. (9.2.1) in components and taking into account the representation (9.2.2) for the electric field, one derives

$$\begin{aligned}
\ddot{x} &= -\frac{e}{m_0} T E_{0x} \sin(k_x x + k_z z) \sum_{n=-\infty}^{\infty} \delta(t - nT) + \omega_0 \dot{y}\,, \\
\ddot{y} &= -\omega_0 \dot{x}\,, \\
\ddot{z} &= -\frac{e}{m_0} T E_{0z} \sin(k_x x + k_z z) \sum_{n=-\infty}^{\infty} \delta(t - nT)\,,
\end{aligned} \tag{9.2.4}$$

where $\omega_0 = eB_0/mc$ is the cyclotron frequency. The second equation in system (9.2.4) can only be integrated once, thus yielding

$$\dot{y} + \omega_0 x = \text{const.} \tag{9.2.5}$$

Using the constant of this motion, the order of the equations of motion (9.2.4) can be reduced and a system of two equations is obtained:

$$\begin{aligned}
\ddot{x} + \omega_0^2 x &= -\frac{e}{m_0} T E_{0x} \sin(k_x x + k_z z) \sum_{n=-\infty}^{\infty} \delta(t - nT) \\
\ddot{z} &= -\frac{e}{m_0} T E_{0x} \sin(k_x x + k_z z) \sum_{n=-\infty}^{\infty} \delta(t - nT)
\end{aligned} \tag{9.2.6}$$

where the constant zero is placed in Eq. (9.2.5). In the case of the oblique propagation of the wave packet ($k_x, k_z \neq 0$), it follows from the equations of motion (9.2.6) that both the longitudinal and transverse degrees of freedom are coupled. The Hamiltonian of system (9.2.6) has the following form:

$$H = \frac{1}{2m_0}(p_x^2 + p_z^2) + \frac{m_0}{2}\omega_0^2 x^2 - e\phi_0 T \cos(k_x x + k_z z) \sum_{n=-\infty}^{\infty} \delta(t - nT) \tag{9.2.7}$$

where $\phi_0 = E_{0x}/k_x = E_{0z}/k_z$ is the amplitude of the potential of the electric field, and $p_x$ and $p_z$ are components of the particle momentum:

$$p_x = m_0 \dot{x}\,, \quad p_z = m_0 \dot{z}\,. \tag{9.2.8}$$

One can replace the system of differential equations in (9.2.6) with a finite-difference system. Between two successive actions of the delta-functions, the trajectory of the particle satisfies the equation

$$\ddot{x} + \omega_0^2 x = 0\,, \quad \ddot{z} = 0\,. \tag{9.2.9}$$

As they pass through the delta function at time $t_n = nT$, the solutions (9.2.9) must satisfy the boundary conditions

$$\begin{aligned} & x(t_n+0) = x(t_n-0)\,, \quad z(t_n+0) = z(t_n-0)\,, \\ & \dot{x}(t_n+0) = \dot{x}(t_n-0) - \frac{e}{m_0} T E_{0x} \sin[k_x x(t_n) + k_z z(t_n)]\,, \\ & \dot{z}(t_n+0) = \dot{z}(t_n-0) - \frac{e}{m_0} T E_{0x} \sin[k_x x(t_n) + k_z z(t_n)]\,. \end{aligned} \tag{9.2.10}$$

Using these equations, one derives from (9.2.6)

$$\begin{aligned} & \dot{x}_{n+1} = -\omega_0 x_n \sin\omega_0 T + \left[\dot{x}_n - \frac{eE_{0x}}{m_0} T \sin(k_x x_n + k_z z_n)\right] \cos\omega_0 T\,, \\ & \dot{x}_{n+1} = x_n \cos\omega_0 T + \left[\dot{x}_n - \frac{eE_{0x}}{m_0} T \sin(k_x x_n + k_z z_n)\right] \sin\omega_0 T\,, \\ & \dot{z}_{n+1} = \dot{z}_n - \frac{eE_{0x}}{m_0} T \sin(k_x x_n + k_z z_n)\,, \quad z_{n+1} = z_n + T\dot{z}_{n+1}\,, \end{aligned} \tag{9.2.11}$$

where the following has been denoted:

$$\dot{x}_n = \dot{x}(nT-0)\,, \quad x_n = x(nT-0)\,, \quad \dot{z}_n = \dot{z}(nT-0)\,, \quad z_n = z(nT-0)\,.$$

By introducing the more convenient dimensionless variables of

$$u_n = k_x \dot{x}_n/\omega_0\,, \quad v_n = -k_x x_n\,, \quad w_n = k_z \dot{z}_n/\omega_0\,, \quad Z_n = k_z z_n\,, \tag{9.2.12}$$

(9.2.11) is rewritten as

$$\begin{aligned} & u_{n+1} = v_n \sin\alpha + [u_n + K\sin(v_n - Z_n)] \cos\alpha\,, \\ & v_{n+1} = v_n \cos\alpha - [u_n + K\sin(v_n - Z_n)] \sin\alpha\,, \\ & w_{n+1} = w_n + K\beta^2 \sin(v_n - Z_n)\,, \quad Z_{n+1} = Z_n + \alpha w_{n+1} \end{aligned} \tag{9.2.13}$$

using the notation

$$\alpha = \omega_0 T\,, \quad K = \frac{eE_{0x}k_x}{m_0\omega_0}T\,. \tag{9.2.14}$$

The 4D-map (9.2.13) obtained corresponds to the coupled web-map and the standard map. To view this, some extreme situations can be considered. If the wave packet propagates along the magnetic field ($E_{0x} = k_x = 0$), the electric field of the packet will not influence the Larmor rotation of the particles in a plane perpendicular to the magnetic field. This means that the longitudinal and transverse degrees of freedom decouple. The system (9.2.13) is reduced to two independent maps. The first of these,

$$u_{n+1} = v_n \sin\alpha + u_n \cos\alpha\,, \quad v_{n+1} = v_n \cos\alpha - u_n \sin\alpha$$

describes the simple rotation of a particle in the magnetic field. The second has the form

$$w_{n+1} = w_n - K\beta^2 \sin Z_n\,, \quad Z_{n+1} = Z_n + \alpha w_{n+1} \tag{9.2.15}$$

and describes only the longitudinal motion along the magnetic field. After substituting $\alpha w = I$, it is reduced to the standard map

$$I_{n+1} = I_n - K_0 \sin Z_n\,, \quad Z_{n+1} = Z_n + I_{n+1} \tag{9.2.16}$$

with the nonlinearity parameter being

$$K_0 = \alpha\beta^2 K = \frac{eE_{0z}k_z}{m_0}T^2\,. \tag{9.2.17}$$

On the other extreme, when the wave packet propagates in a manner that is strictly perpendicular to the magnetic field ($E_{0z} = k_z = 0$), the longitudinal motion is free ($\beta = 0$) while the transverse motion is described by the web-map

$$\begin{aligned} u_{n+1} &= v_n \sin\alpha + (u_n + K\sin v_n)\cos\alpha\,, \\ v_{n+1} &= v_n \cos\alpha - (u_n + K\sin v_n)\sin\alpha\,. \end{aligned} \tag{9.2.18}$$

This map is the generator of the stochastic web with a symmetry of order $q$ if the resonance condition $\alpha = \alpha_q$ is satisfied, where

$$\alpha_q = 2\pi/q \tag{9.2.19}$$

and $q$ is an integer. As mentioned in Section 1.3, condition (9.2.13) refers to the fact that over a full rotation in the magnetic field, a particle experiences exactly $q$ kicks from the wave field. If (9.2.19) is satisfied and the wave packet propagation is strictly transverse to the magnetic field, the phase plane $(u, v)$ in map (9.2.18) is covered by the stochastic web for an arbitrarily small $K$. For the non-orthogonal propagation of the wave packet, different metamorphoses of the phase portrait occur as the parameters of the system change.

Using the dimensionless variables $(u, v; Z, w)$ defined in Eq. (9.2.12), the Hamiltonian (9.2.7) is represented as

$$\begin{aligned}\mathcal{H} &= \mathcal{H}(u, v; Z, w) \\ &= \frac{1}{2}(v^2 + u^2) + \frac{1}{2\beta^2}w^2 - K\cos(v - Z)\sum_{n=-\infty}^{\infty}\delta(\tau - n\alpha)\,,\end{aligned} \tag{9.2.20}$$

where the dimensionless time

$$\tau = \omega_0 t \tag{9.2.21}$$

is introduced. The corresponding Hamiltonian equations of motion are

$$\begin{aligned}\frac{dv}{d\tau} &= -\frac{\partial\mathcal{H}}{\partial u}\,, \quad \frac{du}{d\tau} = \frac{\partial\mathcal{H}}{\partial v}\,, \\ \frac{dw}{d\tau} &= -\beta^2\frac{\partial\mathcal{H}}{\partial Z}\,, \quad \frac{dZ}{d\tau} = \beta^2\frac{\partial\mathcal{H}}{\partial w}\,.\end{aligned} \tag{9.2.22}$$

One can easily verify that system (9.2.22) leads to the map (9.2.13).[1]

## 9.3 Multi-Web Structures in the Phase Space

The conditions for applying the KAM-theory and the Arnold diffusion lead one to conclude that if the unperturbed motion does not have

[1]The material in this section is based on the results of [ZZNSUC 89].

any domain of chaotic motion, or if the domains are very small, a small perturbation therefore guarantees the smallness of the domains of chaos but can change their topology by transferring them into a connected net. A pragmatic formulation of this result can be proposed for the KAM-theory. How difficult is it to find the initial condition for a chaotic orbit? The answer is extremely difficult. In the case where the condition (9.1.1) of the applicability of the KAM-theory is abandoned and the resonant condition (9.2.19) is valid, the answer to the same question is strongly different: it is very easy to find a chaotic orbit in, say, 4D phase space even though the perturbation is very small. The interaction among the different degrees of freedom is strongly enhanced despite the smallness of the coupling constant, a property which will be demonstrated in this section using the 4D-map (9.2.13).[2]

Adopting the same procedure as in Section 8.3, that is, by introducing the rotating frame of reference using the variables

$$u = \rho \cos \phi , \quad v = -\rho \sin \phi , \quad I = \frac{1}{2}\rho^2 \tag{9.3.1}$$

and the generating function

$$F = (\phi - \tau) I \tag{9.3.2}$$

[compare to (8.3.1) and (8.3.2)], the Hamiltonian (9.2.20) is transformed into

$$\tilde{\mathcal{H}} = \mathcal{H} + \frac{\partial F}{\partial \tau} = \frac{1}{2\beta^2} w^2 - K \cos[\rho \sin(\phi + \tau) - Z] \sum_{n=-\infty}^{\infty} \delta(\tau - n\alpha) . \tag{9.3.3}$$

Using the same representation of Eqs. (8.3.4) and (8.3.5) for the sum of $\delta$-functions, Eq. (9.3.3) is transformed into

[2]See footnote on page 199. More results on the same model are found in [NCC 91] and [Ro 97].

$$\tilde{\mathcal{H}} = H_q + V_q$$

$$H_q = \frac{1}{2\beta^2} w^2 - \frac{2K}{\pi q} \sum_{j=1}^{q} \cos\left(v \cos\frac{2\pi}{q} j - u \sin\frac{2\pi}{q} j - Z\right), \tag{9.3.4}$$

$$V_q = -\frac{4}{\pi q} K \sum_{j=1}^{q} \cos\left(v \cos\frac{2\pi}{q} j - u \sin\frac{2\pi}{q} j - Z\right) \sum_{m=1}^{\infty} \cos(m\tau - \alpha m j).$$

The part $H_q$ of the Hamiltonian does not depend on time. Instead, it corresponds to two coupled degrees of freedom. The cases where $q = 1, 2, 4$ are integrable. For example, when $q = 4$:

$$\mathcal{H}_4 = \frac{1}{2\beta^2} w^2 - \frac{1}{\pi} K \cos Z (\cos u + \cos v). \tag{9.3.5}$$

The equations of motion (9.2.22) are

$$\begin{aligned} \frac{dv}{d\tau} &= -\frac{1}{\pi} K \cos Z \sin u, \quad \frac{du}{d\tau} = \frac{1}{\pi} K \cos Z \sin v, \\ \frac{dw}{d\tau} &= -\frac{1}{\pi} \beta^2 K \sin Z (\cos u + \cos v) \\ \frac{dZ}{d\tau} &= w. \end{aligned} \tag{9.3.6}$$

One can conclude from the first pair of equations in (9.3.6) that the quantity

$$(\cos u + \cos v) = \text{const.} = C_0 \tag{9.3.7}$$

is an integral of motion. This motion was described in Section 8.3 [see (8.3.12)]. Therefore, the second pair of Eq. (9.3.6) takes the form

$$\frac{d^2 Z}{d\tau^2} + \frac{1}{\pi} \beta^2 K C_0 \sin Z = 0, \tag{9.3.8}$$

which describes a pendulum with the characteristic frequency of

$$\Omega_z = (\beta^2 K C_0 / \pi)^{1/2}. \tag{9.3.9}$$

The perturbation $V_q$ in (8.3.4) depends periodically on time. It destroys the separatrices of the pendulum (9.3.8) and the $(u, v)$ motion and creates stochastic webs in the 4D-space. The topology of the web is a product of the square lattice in the $(u, v)$-plane and layers in the $(w, Z)$-plane.

In general, $C_0$ is not an integral of motion due to the perturbation $V_q$. Even for a small value of the parameter $\beta$, a slow drift of a particle in the plane $(u, v)$ leads to repeated crossings of the unperturbed separatrices in the $(u, v)$-plane. Such behaviour effectively increases the width of the stochastic web, the diffusion rate along the web, and complicates its topology. The channels of chaotic dynamics for small $\beta$ is called a multi-web.

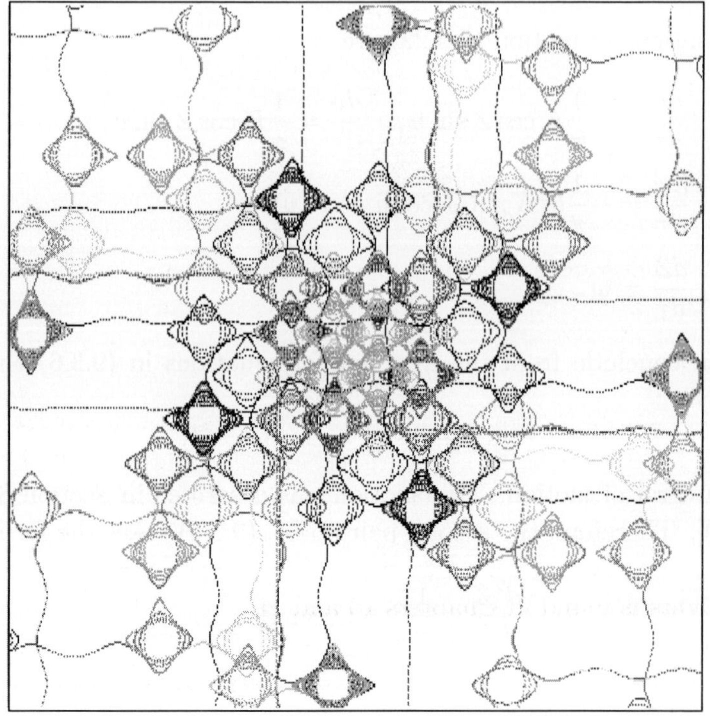

Fig. 9.3.1. Example of fast diffusion in the $(u, v)$ plane for $q = 4$: $K = 0.132$, $\beta^2 = 0.0001$; $w_0 = 1$; the size of the square is $(16\pi)^2$.

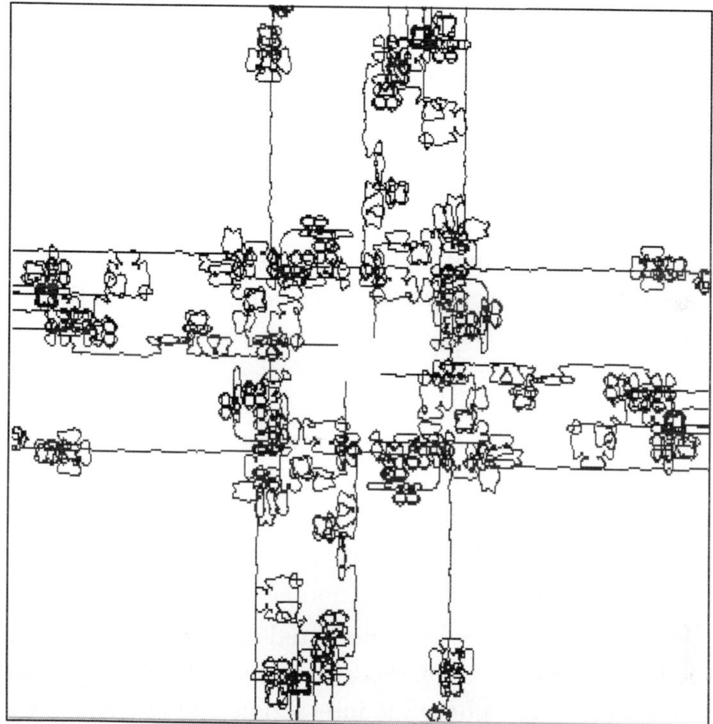

Fig. 9.3.2. Another example with long flights for $q = 4$ and $K = 0.2$, $\beta^2 = 0.004$; $w_0 = 1$; the size of the square is $(40\pi)^2$.

Examples of fast diffusion for very small values of $K$ and $\beta$ are shown in Fig. 9.3.1. Here one can easily observe phenomena such as the trappings of the square web in a cell, escapes from the cell, and "flights" along the web. Sometimes the flights can be enormously long. This is shown in Fig. 9.3.2 and they are called Lévy flights. Further discussion of the flights is found in Chapters 10 and 11.

In cases where $q = 3$ and $q \not\equiv \{q_c\}$, they do not have an integrable Hamiltonian $H_q$ and the diffusion is therefore much faster. At the same time, the diffusion process is highly intermittent due to the presence of different structural elements in the phase space. Examples of chaotic trajectories for $q \not\equiv \{q_c\}$ are presented in the color plates.

## 9.4 Equilibrium of the Atomic Chains

It was mentioned in Section 9.2 that the 4D map (9.2.13) has features resembling some typical physical problems. It was derived for a practical motion in electromagnetic field. This section will demonstrate that the same map describes the equilibrium of an atomic chain with the nearest and next nearest interactions.

When $x_j$ is the co-ordinate of the $j$-th atom and $\Phi(x_1, x_2, \ldots)$ is the potential energy of the atomic chain, the equilibrium conditions are

$$\frac{\partial \Phi}{\partial x_j} = 0\,, \quad (\forall j)\,. \tag{9.4.1}$$

Usually, the chain potential can be presented in a form that indicates the potential energy for each atom:

$$\Phi = \sum_{j=1}^{\infty} \phi_j(x_j, x_{j\pm 1}, \ldots, x_{j\pm k}) \tag{9.4.2}$$

where $\phi_j$ is the $j$-th atom's local potential dependent on the displacement of a finite number, $k$, of the atoms nearest the $j$-th atom. The simplest situation is one where the interaction taking place is with only the nearest particle when $\phi_j$ has the form

$$\phi_j = \Psi(x_j - x_{j-1}) + V(x_j)\,. \tag{9.4.3}$$

Here, $\Psi$ represents the potential energy of interaction with neighbours and $V$ corresponds to a general crystalline field potential at site $j$. In this case, system (9.4.1) is equivalent to a two-dimensional map. Indeed, applying (9.4.1) to (9.4.3), one derives

$$\Psi'_{j+1} = \Psi'_j + V'(x_j) \tag{9.4.4}$$

$$\Psi_j = \Psi(x_j - x_{j-1})\,, \tag{9.4.5}$$

where prime means the derivative with respect to the argument. Using definition (9.4.5) and a new variable,

$$P_j = x_j - x_{j-1}\,, \tag{9.4.6}$$

it follows from (9.4.4) to (9.4.6) that

$$\begin{aligned} P_{j+1} &= (\Psi')^{-1}[\Psi'_j(P_j) + V'(x_j)] \\ x_{j+1} &= x_j + P_{j+1}\,, \end{aligned} \tag{9.4.7}$$

where it is assumed that $\Psi' \neq 0$ anywhere in the domain under consideration and $(\Psi')^{-1}$ refers to the inverse function of $\Psi'$. The iteration Eq. (9.4.7) is equivalent to the initial set of (9.4.1) and correspond to the area-preserving two-dimensional map, which possesses chaotic orbits for some domains of the phase space and some parameter values. In particular, when

$$\Psi_j = \frac{1}{2}(x_j - x_{j-1} - a)^2\,, \tag{9.4.8}$$

(9.4.7) and (9.4.8) yield the map

$$P_{j+1} = P_j + V'(x_j)\,, \quad x_{j+1} = x_j + P_{j+1} \tag{9.4.9}$$

which corresponds to the standard map if $V' = K \sin x_j$. Map (9.4.7) shows that the equilibrium conditions (9.4.1) for the atom's chain can be reformulated as a dynamical system problem. It leads to an increase in the number of degrees of freedom for the equivalent dynamical system if the particles in the chain interact with their next-to-nearest neighbours.[3]

In examining the potential function, $\Phi$, in the form of

$$\begin{aligned} \Phi = \frac{1}{2}\gamma \sum_n (x_{n+1} - x_n - a)^2 - \frac{1}{2}\beta \sum_n (x_{n+2} - x_n - 2a)^2 \\ + V_0 \sum (1 - \cos x_n)\,, \end{aligned} \tag{9.4.10}$$

the first term represents the nearest typical neighbour in the interaction. The second term represents the next-to-nearest neighbour interaction. The last term describes a local potential which can be determined by

[3]The idea of studying the equilibrium of atomic chains by reducing the corresponding conditions to a map and using the methods of dynamical systems was proposed by S. Aubry [Au 79]. Spin-chains were considered in [BTZ 94]. The material in this section was published in [BBZT 84].

a support lattice. Here $\gamma, \beta$ and $V_0$ are some constants of interaction. The equilibrium Eq. (9.4.1) yields

$$-\gamma(x_{n+1} - 2x_n - x_{n-1}) + \beta(x_{n+2} - 2x_n + x_{n-2}) + V_0 \sin x_n = 0 \,. \tag{9.4.11}$$

Equation (9.4.11) can be reduced to a four-dimensional map similar to (9.2.13). To do so, one introduces the new variables of

$$I_{n+1} = x_{n+1} - x_n \,, \quad \xi_{n+1} = \eta_{n+1} - \eta_n \,, \quad \eta_n = I_{n+1} - I_n \tag{9.4.12}$$

and put some restrictions on the constants $\beta$ and $\gamma$, namely, $\gamma < 0$, $\beta < 0$ and $|\gamma| < 4|\beta|$. Denote

$$\omega_0^2 = 4(1 - \gamma/4\beta) \equiv 4\sin^2 \alpha/2 \,. \tag{9.4.13}$$

Using Eqs. (9.4.12) and (9.4.13), the main equilibrium Eq. (9.4.11) can be rewritten as

$$\xi_{n+1} = \xi_n - \omega_0^2 \eta_n - (V_0/\beta) \sin x_n \,. \tag{9.4.14}$$

The next transform of the variables

$$\begin{aligned} w_n &= I_n + \bar{\xi}_n \,, \quad & z_n &= x_n + \bar{\eta}_n \\ \bar{\xi} &= \xi/\omega_0^2 \,, & \bar{\eta} &= \eta/\omega_0^2 \end{aligned} \tag{9.4.15}$$

and Eqs. (9.4.14) and (9.4.12) lead to the map

$$\begin{aligned} \bar{\xi}_{n+1} &= \bar{\xi}_n - \omega_0^2 \bar{\eta}_n - (V_0/\omega_0^2 \beta) \sin(z_n - \bar{\eta}_n) \\ \bar{\eta}_{n+1} &= \bar{\eta}_n + \bar{\xi}_{n+1} \\ w_{n+1} &= w_n - (V_0/\omega_0^2 \beta) \sin(z_n - \bar{\eta}_n) \\ z_{n+1} &= z_n + w_{n+1} \,. \end{aligned} \tag{9.4.16}$$

When a rotational transform for the $(\bar{\xi}, \bar{\eta})$ variables is performed:

$$\bar{\xi} = v(1 - \cos \alpha) - u \sin \alpha \,, \quad \bar{\eta} = v \tag{9.4.17}$$

which gives

$$\begin{aligned} u_{n+1} &= (u_n + K\sin(v_n - z_n))\cos\alpha + v_n \sin\alpha \\ v_{n+1} &= -(u_n + K\sin(v_n - z_n))\sin\alpha + v_n \cos\alpha \\ w_{n+1} &= w_n - K\sin\alpha \cdot \sin(v_n - z_n) \\ z_{n+1} &= z_n + w_{n+1} \end{aligned} \tag{9.4.18}$$

with the potential parameter being

$$K \equiv \frac{|V_0|}{|\beta|\omega_0^2 \sin\alpha}, \quad V_0 < 0. \tag{9.4.19}$$

The parameter $K$ is a combination of the one-site potential amplitude $|V_0|$ and next-to-nearest interaction constant $\beta$. The new system, (9.4.18), coincides with map (9.2.13) which describes a particle motion in a constant magnetic field and in a wave packet. Certain formal reasons exist to account for the coincidence which is a result of the choice of the potential energy $\phi$ in the form (9.4.10).

Hence the problem of atomic chain equilibrium is reduced to a known problem and one can now describe some of its properties. The first pair of equations, (9.4.18), is known as the web-map. The second pair is known as the standard map. The most interesting cases correspond to the resonance situation when

$$\alpha = \alpha_q = 2\pi/q \tag{9.4.20}$$

and $q$ is an integer (a more general resonance case is when $\alpha/2\pi$ is rational). The resonance condition (9.4.20) means that a special relationship exists between parameters $\gamma$ and $\beta$ [see Eq. (9.4.13)]. In particular, for $q = 4$, one has $\omega_0^2 = 1/2$ and $\gamma = 2\beta$. For large $q$, the parameter $\alpha$ is small and $\gamma \approx 4\beta(1 - \pi^2/q^2)$. The infinite diffusion growth of $u$ and $v$ occurs if an initial condition, $(u_0, v_0, w_0, z_0)$, is taken inside the web. At the same time, no growth in $w$ is found for small $K$. From the definitions (9.4.12) and (9.4.15), such instability leads to a disordering of the particle chain as a whole. In spite of the diffusion disordering process, there is a non-trival possibility that order may exist in some parts of the chain. As shown in Section 9.3 for the map (9.2.13), sporadic and intermittent, almost regular parts of orbits generated by the map exist. These parts are called "flights" and can be extremely long. They occur as a result of a non-trivial topology in the phase space of the system (9.4.18). The flights correspond to almost ordered parts of the chain.

The case of $K \gtrsim 1$, where there is strong chaos and strong particle disordering, will not be discussed here.

## 9.5 Discretisation

Sometimes, a differential equation is replaced by a corresponding difference equation or vice versa. In contrast to the initial differential equation, difference equations have an additional half degree of freedom and consist of a time-dependent force. It is generally stated that the discretisation procedure creates stochastic layers or webs in the phase space of the system. This statement is borne out by the following examples.

When a pendulum equation such as

$$\ddot{x} + \omega_0^2 \sin x = 0 \tag{9.5.1}$$

is replaced by its discrete version,

$$x_{n+1} - 2x_n + x_{n-1} + \omega_0^2 (\Delta t)^2 \sin x_n = 0\,, \tag{9.5.2}$$

where $\Delta t$ is the time-step of discretisation and

$$x_n = x(n\Delta t)\,, \tag{9.5.3}$$

the step is usually small and the following inequality is required:

$$K \equiv (\omega_0 \Delta t)^2 \ll 1\,. \tag{9.5.4}$$

Using the notation

$$I_n = x_n - x_{n-1} = \Delta t \cdot p_n\,, \tag{9.5.5}$$

Eqs. (9.5.2) and (9.5.5) are rewritten as

$$\begin{aligned} I_{n+1} &= I_n - K \sin x_n \\ x_{n+1} &= x_n + I_{n+1} \end{aligned} , \tag{9.5.6}$$

which coincides with the standard map. The standard map has a different topology from the pendulum (9.5.1). Even for an arbitrarily small $K$, stochastic layers and resonance island chains exist (see Sections 1.2 and 2.7).

The Hamiltonian of the pendulum is

$$H_0 = \frac{1}{2}p^2 - \omega_0^2 \cos x \tag{9.5.7}$$

(see Section 1.4) whereas that for the standard map (9.5.6) is

$$H = \frac{1}{2}p^2 - \omega_0^2 \cos x \sum_{m=-\infty}^{\infty} \cos m\nu t\,, \quad \nu = \frac{2\pi}{\Delta t} \tag{9.5.8}$$

(see Section 1.2). One can write

$$H = H_0 + V_{\text{discr}} \tag{9.5.9}$$

and, comparing (9.5.7) and (9.5.8), derive

$$V_{\text{discr}} = -2\omega_0^2 \cos x \cdot \cos \nu t \tag{9.5.10}$$

up to the higher order terms in frequency $\nu$. The potential $V_{\text{discr}}$ corresponds to the force originating from the discretisation procedure. It has high frequency oscillations,

$$\nu \gg \omega_0\,, \tag{9.5.11}$$

due to the condition (9.5.4). After averaging over high frequency oscillations, the corrections to the original (unperturbed) solutions will be of the order $K$, that is, small. Nevertheless, when the averaging method is applied, the effects of the topological modification of the phase space and the appearance of stochastic layers are missing.

Turning now to a more complicated example of a perturbed oscillator,

$$\ddot{x} + \omega_0^2 x = \epsilon \sin(kx + \nu t) \tag{9.5.12}$$

(see Sections 1.5 and 4.2), the difference equation is

$$x_{n+1} - 2x_n + x_{n-1} + \omega_0^2 \Delta t^2 x_n = \epsilon(\Delta t)^2 \sin(kx_n + n\nu\Delta t)\,. \tag{9.5.13}$$

Using notation (9.5.5), (9.5.13) is rewritten as a map:

$$\begin{aligned} I_{n+1} &= I_n - (\omega_0\Delta t)^2 x_n + \epsilon(\Delta t)^2 \sin(kx_n + n\nu\Delta t) \\ x_{n+1} &= x_n + I_{n+1}\,. \end{aligned} \tag{9.5.14}$$

If $\omega_0 = \nu = 0$, (9.5.14) is reduced to the standard map (9.5.6). Replacing the variables

$$x_n = v_n\,, \quad I_n = -\sin\alpha \cdot u_n + (1 - \cos\alpha)v_n \tag{9.5.15}$$

where

$$(\omega_0\Delta t)^2 = 4\sin^2\alpha/2 \tag{9.5.16}$$

is placed, map (9.5.14) in $(u, v)$-variables is

$$\begin{aligned} u_{n+1} &= (u_n - K_0 \sin(kv_n + n\nu\Delta t))\cos\alpha + v_n \sin\alpha \\ v_{n+1} &= -(u_n - K_0 \sin(kv_n + n\nu\Delta t))\sin\alpha + v_n \cos\alpha \end{aligned} \tag{9.5.17}$$

with

$$K_0 = \epsilon(\Delta t)^2/\sin\alpha\,. \tag{9.5.18}$$

For $\nu = 0$ and $k = 1$, map (9.5.17) coincides with the web-map (see Section 1.5) and generates a stochastic web in the phase space when $\alpha$ is close to the resonant condition

$$\alpha_q = 2\pi/q \tag{9.5.19}$$

with integer $q$. At the same time, Eq. (9.5.12) is integrable for $\nu = 0$.

The discretisation version of Eq. (9.5.12) can be formulated as a problem of the atomic chain equilibrium. For this purpose, consider the potential function of atoms in the form of

$$\Phi = \frac{1}{2}\gamma \sum_n (x_{n+1} - x_n)^2 + \frac{1}{2}\omega_0^2 \sum_n x_n^2 + V_0 \sum_n \cos k(x_n + na)\,, \tag{9.5.20}$$

which is somewhat different from (9.4.10). Expression (9.5.20) does not include next-to-nearest term of the potential at the site of the atom. The equilibrium conditions (9.4.1) lead to the map

$$\begin{aligned} I_{n+1} &= I_n - (\omega_0^2/\gamma)x_n + (V_0 k/\gamma)\sin(kx_n + na) \\ x_{n+1} &= x_n + I_{n+1}\,. \end{aligned} \tag{9.5.21}$$

Map (9.5.21) coincides with (9.5.14) up to the notations of the constants. Therefore, map (9.5.21) can be rewritten in the same form as (9.5.17) if the notations

$$K_0 = kV_0/\gamma \sin\alpha\,, \quad \nu\Delta t = ka\,, \quad \omega_0^2/\gamma = 4\sin^2\alpha/2 \tag{9.5.22}$$

are used.

The results obtained show that the time discretisation of the initial differential equation(s) adds an additional 1/2 degree of freedom to the original system. The new and different type of equations can coincide with the equations of equilibrium of the particle chain.[4]

## Conclusions

1. The coupled web-map and the standard map describe a charged particle motion in a constant magnetic field and oblique electromagnetic wave packet. A four-dimensional map corresponds to a system with 2 1/2 degrees of freedom in order that the Arnold diffusion can take place. In fact, a much faster diffusion exists in a non-exponentially

[4]More information on the analogy between discretised differential equations and the equilibrium of atomic chains can be found in [BTZ 94].

small domain because of the presence of degeneracy. One can always expect fast non-Arnold diffusion to take place for at least one degree of freedom which corresponds to a linear oscillator in the absence of coupling.

2. The equilibrium condition for a one-dimensional atomic chain can be written in the form of a map. The order of the map is the same as the number of nearest interacting atoms.

*Chapter 10*

# NORMAL AND ANOMALOUS KINETICS

## 10.1 Fokker-Planck-Kolmogorov (FPK) Equation

The idea of using a so-called kinetic equation is based on a simplifed description of dynamics at the expense of loss of information about trajectories. A typical approach is to consider the distribution function $F = F(I, \vartheta, t)$ in the $2N$-dimensional phase space and to apply the technique of coarse-graining which means averaging over some fast variables such as phases $\vartheta$:

$$P(I,t) \equiv \langle\langle F(I,\vartheta,t)\rangle\rangle = \frac{1}{2\pi}\int_0^{2\pi} d\vartheta F(I,\vartheta,t) \qquad (10.1.1)$$

where $d\vartheta = d\vartheta_1 \dots d\vartheta_N$. One reason for using coarse-graining is due to the existence of at least two different time scales: $t_c$, which is a scale for the averaging (10.1.1) and $t_d$, a scale of slow evolution of $P(I,t)$, and

$$t_c \ll t_d \,. \qquad (10.1.2)$$

An equation that describes the slow evolution of $P(I,t)$ is called a kinetic equation and an equation of evolution of moments of $P(I,t)$, that is,

$$\langle I^n \rangle = \int dI\, I^n\, P(I,t)\,, \qquad (10.1.3)$$

is called a transport equation. Chaotic dynamics provide a possibility of introducing appropriate coarse-graining. The main thrust of this chapter and the next is to find the corresponding kinetic and transport equations. Towards that end, one begins with a consideration of Fokker-Planck-Kolmogorov (FPK) equation modified from the original idea by A. Kolmogorov.[1]

When $x$ is a characteristic variable of the wandering process, such as a particle co-ordinate, and $P(x,t)$ is the probability density of having the particle co-ordinate $x$ at an instant $t$, the Chapman-Kolmogorov equation is

$$W(x_3, t_3|x_1, t_1) = \int dx_2 W(x_3, t_3|x_2, t_2) \cdot W(x_2 t_2|x_1, t_1), \qquad (10.1.4)$$

where $W(x, t|x', t')$ is the probability density of having the particle at position $x$ and at time $t$ if it was at point $x'$ at time $t' \leq t$. The connection between $P$ and $W$ is given below. A typical assumption,

$$W(x, t|x', t') = W(x, x'; t - t'), \qquad (10.1.5)$$

corresponds to the regular kinetic equation derivation scheme. For small $\Delta t = t - t'$, one has an expansion,

$$W(x, x_0; t + \Delta t) = W(x, x_0; t) + \Delta t \frac{\partial W(x, x_0; t)}{\partial t}, \qquad (10.1.6)$$

where the higher order terms are neglected. Equation (10.1.6) provides the existence of the limit

$$\lim_{\Delta t \to 0} \frac{1}{\Delta t} \{W(x, x_0; t + \Delta t) - W(x, x_0; t)\} = \frac{\partial W(x, x_0; t)}{\partial t}. \qquad (10.1.7)$$

Equation (10.1.4) consists of only one function, $W$, with different arguments. The main idea of the kinetic equation derivation is to distinguish

---

[1] In his original publication [K 38], A. Kolmogorov formulated the formal conditions leading to the FPK-equation of a general type. The main condition is the existence of the second moment, $\langle\langle(\Delta x)^2\rangle\rangle/\Delta t$, for the mean local change per unit time of the square displacement, $\Delta x$, in the limit, $\Delta t \to 0$, and the vanishing of all higher moments of $\Delta x$. L. Landau [L 37] also derived the diffusion-type equation which indicates that the detailed balance principle leads to a special type of the FPK-equation that preserves the number of particles. In the derivation presented here (see also [Za 94-3]), a modified Kolmogorov's scheme giving directly the FPK equation in the Landau form is used.

those functions $W$ that correspond to different time scales. One now uses

$$t_3 = t + \Delta t\,, \quad t_1 = t\,, \quad t_2 = t_3 - t_1 = \Delta t \tag{10.1.8}$$

and the notation

$$P(x,t) \equiv W(x, x_0; t)\,. \tag{10.1.9}$$

Equations (10.1.8) and (10.1.9) reinforce the typical assumption that only large time-scale asymptotics,

$$t \gg t_c\,, \tag{10.1.10}$$

should be considered, that is, large deviations $|x - x_0|$ are expected. The value $t_c$ in (10.1.10) is the characteristic time taken for "interaction", "phase mixing" or something else which defines the short time scale of the process. This is usually a characteristic time over which coarse-graining should be performed. Using (10.1.4), (10.1.8) and (10.1.9), Eq. (10.1.7) is rewritten as

$$\frac{\partial P(x,t)}{\partial t} = \lim_{\Delta t \to 0} \frac{1}{\Delta t} \int dy \{W(x, y; \Delta t) - \delta(x - y)\} P(y, t)\,. \tag{10.1.11}$$

It is the existence of $t_c$ that makes a difference between $P = W(x, y; t)$ in (10.1.9) and the kernel function $W(x, y; \Delta t)$ since $t \to \infty$ and $\Delta t \sim t_c$ are finite. In other words, the limit ($\Delta t \to 0$, $t$-finite) is considered in the same sense as the limit ($\Delta t$-finite, $t \to \infty$). The initial condition for $W(x, y, \Delta t)$ can be taken in an obvious form as

$$\lim_{\Delta t \to 0} W(x, y; \Delta t) = \delta(x - y)\,. \tag{10.1.12}$$

For a small but non-zero $\Delta t$, one can write the expansion, over derivatives, of the $\delta$-function as

$$W(x, y; \Delta t) = \delta(x - y) + A(y; \Delta t)\delta'(x - y) + \frac{1}{2} B(y; \Delta t)\delta''(x - y)\,, \tag{10.1.13}$$

where $A$ and $B$ are some functions and prime means derivative with respect to the argument. The functions $A, B$ can be expressed through the moments of the transition probability $W(x, y; \Delta t)$:

$$\begin{aligned} A(y;\Delta t) &= \int_{-\infty}^{\infty} dx(y-x)W(x,y;\Delta t) \equiv \langle\langle \Delta y \rangle\rangle \\ B(y;\Delta t) &= \int_{-\infty}^{\infty} dx(y-x)^2 W(x,y;\Delta t) \equiv \langle\langle (\Delta y)^2 \rangle\rangle \,. \end{aligned} \tag{10.1.14}$$

That is how coarse-graining is performed.

The transition probability, $W(x, y; \Delta t)$, should satisfy the normalisation conditions:

$$\begin{aligned} \int dx W(x,y;\Delta t) &= 1 \\ \int dy W(x,y;\Delta t) &= 1 \end{aligned} \,. \tag{10.1.15}$$

After substituting (10.1.13) and (10.1.14) in (10.1.15), one obtains either

$$A(y;\Delta t) = \frac{1}{2}\frac{\partial B(y;\Delta t)}{\partial y} \tag{10.1.16}$$

or

$$\langle\langle \Delta y \rangle\rangle = \frac{1}{2}\frac{\partial}{\partial y}\langle\langle (\Delta y)^2 \rangle\rangle \,. \tag{10.1.17}$$

Equations (10.1.16) and (10.1.17) were obtained by Landau using the detailed balance principle but in the scheme of this book, they come about automatically (see footnote [1]).

Assuming that the limits

$$\begin{aligned} \lim_{\Delta t \to 0} \frac{1}{\Delta t}\langle\langle (\Delta x) \rangle\rangle &\equiv \mathcal{A}(x) \\ \lim_{\Delta t \to 0} \frac{1}{\Delta t}\langle\langle (\Delta x)^2 \rangle\rangle &\equiv \mathcal{B}(x) \end{aligned} \tag{10.1.18}$$

exist, the FPK equation is obtained by substituting (10.1.13) in (10.1.11):

$$\frac{\partial P(x,t)}{\partial t} = -\frac{\partial}{\partial x}(\mathcal{A}(x)P(x,t)) + \frac{1}{2}\frac{\partial^2}{\partial x^2}(\mathcal{B}(x)P(x,t)) \,. \tag{10.1.19}$$

It can be rewritten in a divergence-like form,

$$\frac{\partial P}{\partial t} = \frac{1}{2}\frac{\partial}{\partial x}\left(\mathcal{B}\frac{\partial P}{\partial x}\right) \tag{10.1.20}$$

after applying the relation,

$$\mathcal{A}(x) = \frac{1}{2}\frac{\partial \mathcal{B}(x)}{\partial x}, \tag{10.1.21}$$

which follows from (10.1.17) and (10.1.18).

The existence of limits (10.1.18) is crucial to the derivation of the FPK equation. These limits define the local structure of the trajectories' behaviour in small time and space domains. It has already been shown that fractal local properties can be an alternative to regular expansions such as (10.1.6) or (10.1.13). An example is considered in Chapter 11.

The generalisation of the FPK equation in the case of a vector variable $\mathbf{x}$ is fairly simple:

$$\frac{\partial P}{\partial t} = \frac{1}{2}\sum_{j,k}\frac{\partial}{\partial x_j}\mathcal{B}_{jk}\frac{\partial P}{\partial x_k}. \tag{10.1.22}$$

Here the relation

$$\mathcal{A}_j(x) = \frac{1}{2}\sum_k \frac{\partial \mathcal{B}_{jk}}{\partial x_k} \tag{10.1.23}$$

and notations

$$\begin{aligned} \mathcal{A}_j(x) &= \lim_{\Delta t \to 0}\frac{1}{\Delta t}\langle\langle \Delta x_j \rangle\rangle \\ \mathcal{B}_{jk}(x) &= \lim_{\Delta t \to 0}\frac{1}{\Delta t}\langle\langle \Delta x_j \Delta x_k \rangle\rangle \end{aligned} \tag{10.1.24}$$

are used.

For a simplified situation where $\mathcal{B} = \text{const.}$, Eq. (10.1.20) yields

$$\frac{\partial P}{\partial t} = \frac{1}{2}\mathcal{B}\frac{\partial^2 P}{\partial x^2} \tag{10.1.25}$$

and the constant $\mathcal{B}$ is termed a diffusion coefficient. The simple result,

$$\langle x^2 \rangle = \mathcal{B}t\,, \quad t \to \infty \tag{10.1.26}$$

defines the asymptotic behaviour of the second moment of $P$ as time becomes fairly large. The behaviour (10.1.25) and (10.1.26) is called by one of three names: diffusion, normal diffusion and normal transport.

## 10.2 Transport for the Standard Map and Web-Map

The FPK equation can be applied to some typical maps. For the standard map (1.28),

$$p_{n+1} = p_n - K \sin x_n\,, \quad x_{n+1} = x_n + p_{n+1} \tag{10.2.1}$$

and $x$ is the fast variable. For $K \gg 1$, there is a strong mixing process consisting of phases and their distribution is close to being uniform. Therefore,

$$\langle\langle \sin x \rangle\rangle = 0\,, \quad \langle\langle \sin^2 x \rangle\rangle = 1/2 \tag{10.2.2}$$

where $\langle\langle \cdots \rangle\rangle$ means averaging over the phase. Thus, the simplest consideration yields

$$\begin{aligned} \langle\langle \Delta p \rangle\rangle &\equiv \langle\langle p_{n+1} - p_n \rangle\rangle = -K \langle\langle \sin x \rangle\rangle = 0 \\ \langle\langle \Delta p^2 \rangle\rangle &\equiv \langle\langle (p_{n+1} - p_n)^2 \rangle\rangle = K^2 \langle\langle \sin^2 x \rangle\rangle = \frac{1}{2} K^2\,, \end{aligned} \tag{10.2.3}$$

where (10.2.2) is used. Turning now to $P = P(p, t)$, where $T$ is the interval between two adjacent steps, it follows from definition (10.1.18) that

$$\mathcal{B}(p) = \text{const.} = K^2/2T\,, \tag{10.2.4}$$

and that from (10.1.26) the transport is defined by the asymptotic law:

$$\langle p^2 \rangle = \mathcal{B}t = \frac{1}{2} K^2 \frac{t}{T}\,. \tag{10.2.5}$$

A necessary condition for the FPK equation to become valid is the smallness of the change for each step:

$$\langle\langle (\Delta p)^2 \rangle\rangle \ll \langle p^2 \rangle\,. \tag{10.2.6}$$

It is clear from (10.2.3) and (10.2.5) that this simply means a fairly large time, that is,

$$t \gg T\,. \tag{10.2.7}$$

In the limit $\Delta t \to 0$, $\Delta t = T$ and is fixed. Instead of $\Delta t \to 0$, $t/\Delta t \to \infty$ is considered which coincides with (10.2.7). In this way, the FPK equation can be applied to a discrete map.

The case of $K \gg 1$ is again considered in the diffusion of the web-map (1.3.5),

$$\hat{M}_\alpha\colon \begin{array}{l} u_{n+1} = (u_n + K\sin v_n)\cos\alpha + v_n \sin\alpha \\ v_{n+1} = -(u_n + K\sin v_n)\sin\alpha + v_n\cos\alpha \end{array}. \tag{10.2.8}$$

The same result, (10.2.5), is obtained for the map $\hat{M}_\alpha$:

$$\langle R^2 \rangle = \frac{1}{2} K^2 \frac{t}{T} \tag{10.2.9}$$

where the slow variable is the distance in the phase space,

$$R = (u^2 + v^2)^{1/2}\,, \tag{10.2.10}$$

and the diffusion coefficient $\mathcal{B}(R) = \text{const.}$ and coincides with (10.2.4).

What is the behaviour of the moments really like? The answer to this question follows from Fig. 10.2.1 for the standard map and Fig. 10.2.2 for the web-map of the four-fold symmetry, that is,

$$u_{n+1} = v_n\,, \quad v_{n+1} = -u_n - K\sin v_n\,. \tag{10.2.11}$$

There are very strong oscillations near the value (10.2.4), and even for some values of $K$, a behaviour resembling singularity takes place. The oscillations of the diffusion constant as a function of $K$ have been known for some time, and some analytical expressions were even derived via a partial inclusion of the phase correlations for adjacent steps.[2]

---

[2]The oscillations in the coefficient of diffusion as a function of the parameter $K$ for the standard map were discovered in numerical simulation [C 79]. The theoretical explanation was developed in [RW 80] and then improved by including the influence of the so-called cantori [CM 81, DMP 89]. The corresponding formula for the diffusion oscillation of the web-map was derived in [ACSZ 90] and [DA 95].

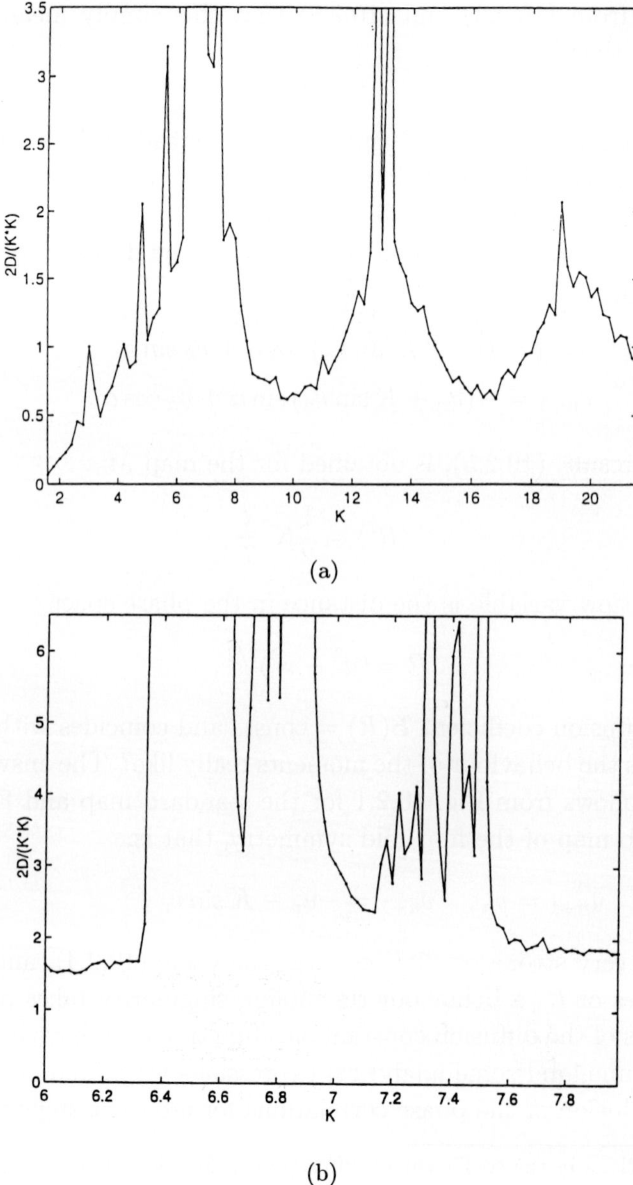


Fig. 10.2.1. Normalised diffusion coefficient for the standard map versus parameter $K$: (a) low resolution plot; (b) a magnified part of (a); and (c) higher resolution plot for the same interval as in (b).

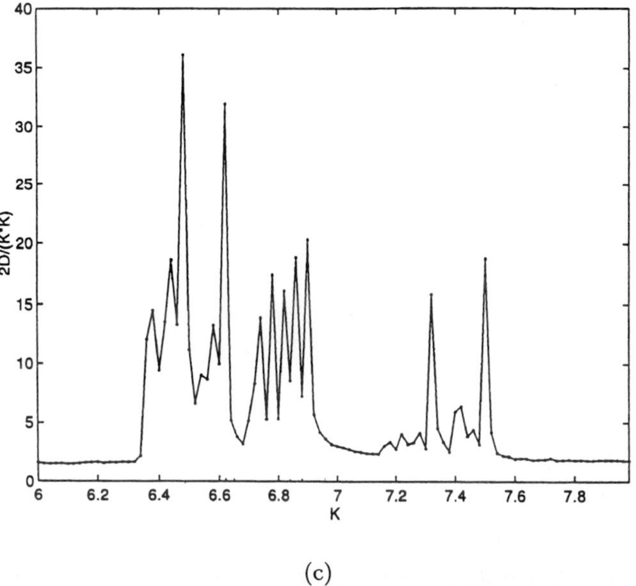


(c)

Fig. 10.2.1. (*Continued*)

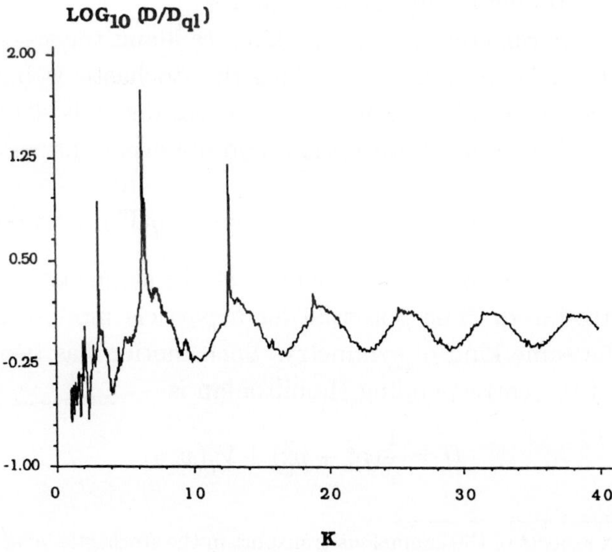


Fig. 10.2.2. Normalised diffusion coefficient for the web-map versus the parameter $k$.

The peaks in Figs. 10.2.1 and 10.2.2 are the result of a non-diffusional character in the evolution of the moments and, strictly speaking, of the absence of the FPK-type equation. The simulation shows that

$$\langle R^2 \rangle \sim t^{\mu_R}, \quad \langle p^2 \rangle \sim t^{\mu_p} \tag{10.2.12}$$

with exponents $\mu_R, \mu_p$ satisfying the conditions

$$1 \leq \mu_R, \mu_p < 2, \tag{10.2.13}$$

which depend on the values of the parameter $K$. The behaviour of the second moments, defined by (10.2.12) and (10.2.13), is called either anomalous transport, superdiffusion, or anomalous diffusion. In some situations, it is possible that the corresponding exponent $\mu < 1$, that is, a subdiffusion, exists. Sometimes the process of mixing in the phase space is less strong or uniform than is assumed during the derivation of the kinetic equation of the FPK-type. One has already seen that in the phase space, fractal quasi-traps and other types of singularities can occur which either enhance or suppress the diffusional process. In the next chapter, this issue is discussed in greater detail. This chapter will restrict itself to illustrating a few examples.

When the parameter $K < 1$, the kinetics along the stochastic layer for the standard map (along $x$) or along the stochastic web are derived. Again, the character of the diffusional process depends strongly on the value of $K$ as it displays both normal and abnormal diffusion.[3]

## 10.3 Dynamics in the Potential with $q$-Fold Symmetry

In Section 8.3, a Hamiltonian having a $q$-fold symmetry in the phase space is introduced. The potential for a particle motion in 2D space can have the same kind of symmetry. Such motion has two degrees of freedom and the corresponding Hamiltonian is

$$H = \frac{1}{2}(p_x^2 + p_y^2) + V_q(x, y) \tag{10.3.1}$$

[3]The different aspects of the anomalous transport in the stochastic layer were studied in [Ka 83], [BZ 83], [CS 84] and [Ba 97]. Comments on the high rate of diffusion through a stochastic web as compared to the Arnold diffusion are found in [LW 89].

with the potential

$$V_q = \sum_{j=1}^{q} \cos(k\mathbf{r} \cdot \mathbf{e}_j), \tag{10.3.2}$$

where $k = 2\pi/\ell$ defines a characteristic length scale $\ell$ and $\mathbf{r} = (x, y)$ and $\mathbf{e}_j$ are unit vectors that form a regular star:

$$\mathbf{e}_j = [\cos(2\pi j/q), \sin(2\pi j/q)]. \tag{10.3.3}$$

For $q = 4$, the potential

$$V_4 = 2(\cos kx + \cos ky) \tag{10.3.4}$$

defines the square lattice symmetry and the corresponding dynamical problem,

$$\dot{p}_x = -\frac{\partial H}{\partial x}, \quad \dot{p}_y = -\frac{\partial H}{\partial y}, \quad \dot{x} = \frac{\partial H}{\partial p_x}, \quad \dot{y} = \frac{\partial H}{\partial p_y}, \tag{10.3.5}$$

is integrable.

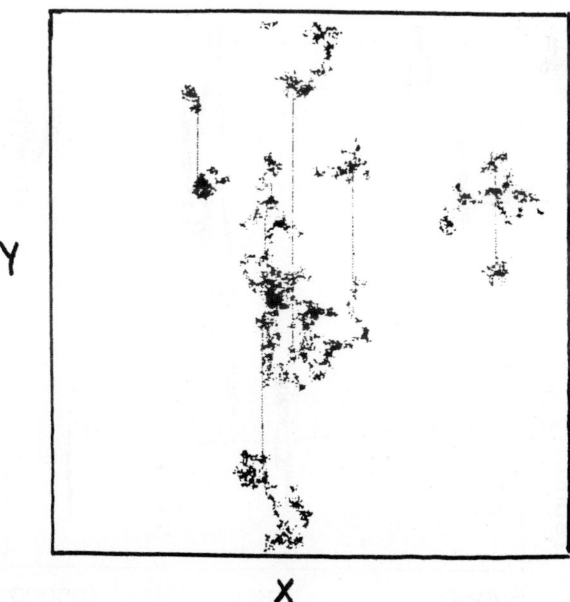


Fig. 10.3.1. Poincaré section of the chaotic random walk in the hexagonal potential with long flights: $E = 2.0$; square size is $(1400\pi)^2$.

The problem (10.3.5) becomes non-integrable if one considers a perturbation in (10.3.4) which destroys the symmetry or when $q \neq 1, 2, 4$. The potential for $q = 3$ is still periodic. However, all the other potentials with $q \not\equiv \{q_c\}$ possess a quasi-crystal type of symmetry. A particle performs a kind of random walk in space which can be characterised by the transport exponent $\mu$:

$$\langle R^2 \rangle \sim t^{\mu} \tag{10.3.6}$$

where $t \to \infty$ and $R^2 = x^2 + y^2$. At some intervals of energy $E = H$, stochastic webs are found where the random walk process can occur and, generally speaking, the processes correspond to the anomalous transport rather than the normal one. Some examples of a random walk trajectory are shown in Fig. 10.3.1 for the kagome-lattice potential

$$V_3 = \cos x + \cos\left(\frac{1}{2}x + \frac{3^{1/2}}{2}y\right) + \cos\left(\frac{1}{2}x + \frac{3^{1/2}}{2}y\right), \tag{10.3.7}$$

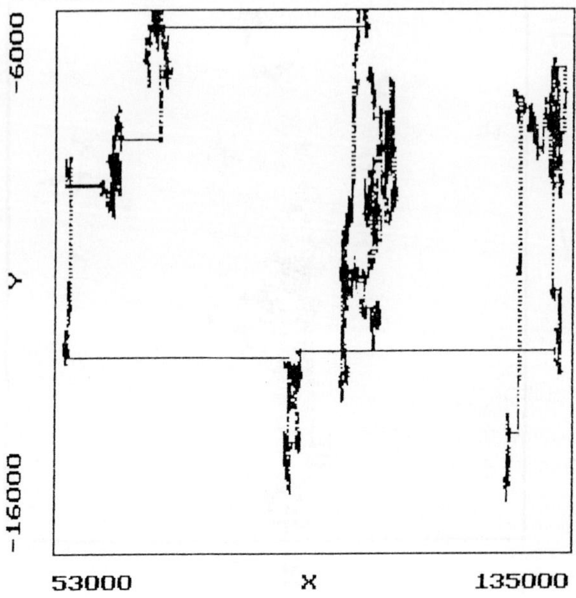


Fig. 10.3.2. The same as in Fig. 10.3.1 but for the perturbed square potential and $E = 1.5$.

and in Fig. 10.3.2 for the perturbed square-symmetry potential

$$V(x,y) = \cos x + \cos y + \epsilon \cos x \cos y\,. \tag{10.3.8}$$

In both examples, $\mu > 1$ and extremely long flights are observed which indicate the existence of memory in a fairly large part of the trajectories in spite of their random behaviour.[4]

## 10.4 More Examples of the Anomalous Transport

Like the previous section, this section aims to demonstrate an unusual random walk performance for a system with dynamical chaos.

The first example is related to the standard map. The trajectory is shown in Fig. 10.4.1 for

$$K = 6.908745\,, \tag{10.4.1}$$

which is in the vicinity of the accelerator mode threshold $K_c = 2\pi = 6.28\ldots$. One can observe enormously long flights which lead to the anomalous transport with exponents

$$\mu_p = 1.25\,, \quad \mu_x = 3.3\,. \tag{10.4.2}$$

The definitions of the exponents are from

$$\langle p^2 \rangle \sim t^{\mu_p}\,, \quad \langle x^2 \rangle \sim t^{\mu_x} \tag{10.4.3}$$

and the trivial connection between $\mu_x$ and $\mu_p$ is

$$\mu_x = \mu_p + 2\,. \tag{10.4.4}$$

The data in (10.4.3) are in agreement with (10.4.4). (Compare (10.4.1) to (3.5.16) and Fig. 3.5.2).[5]

Thc second example pertains to the web-map with a four-fold symmetry (10.2.11) and

$$K = 6.349972\,. \tag{10.4.5}$$

---

[4]For more information on the anomalous diffusion in two-dimensional lattice potential, see [GNZ 84], [GZR 87], [ZSCC 89] and [CZ 91]. Figures 10.3.1 and 10.3.2 are taken from the last two references.

[5]The data for Figs. 10.4.1 and 10.4.2 are taken from [ZEN 97] and [ZN 97].

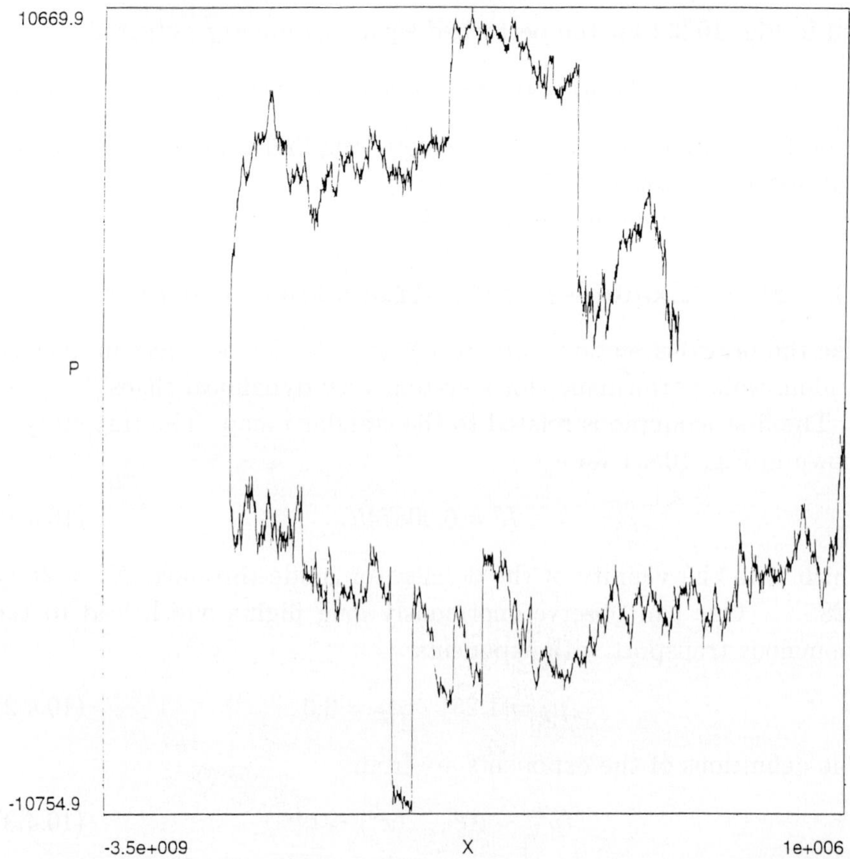


Fig. 10.4.1. A typical trajectory with parabolic flights for the standard map and $K = 6.908745$ near the threshold of the accelerator mode.

(Compare this to (3.5.8) and Fig. 3.5.1). A typical trajectory with long flights is presented in Fig. 10.4.2. The corresponding value of $\mu$ in (10.3.6) is

$$\mu = 1.26 . \tag{10.4.6}$$

In summary, one should note that the stickiness in the islands' boundary induces anomalous transport. In fact, it can be attributed to many other causes resulting from a non-uniformity in the phase space.

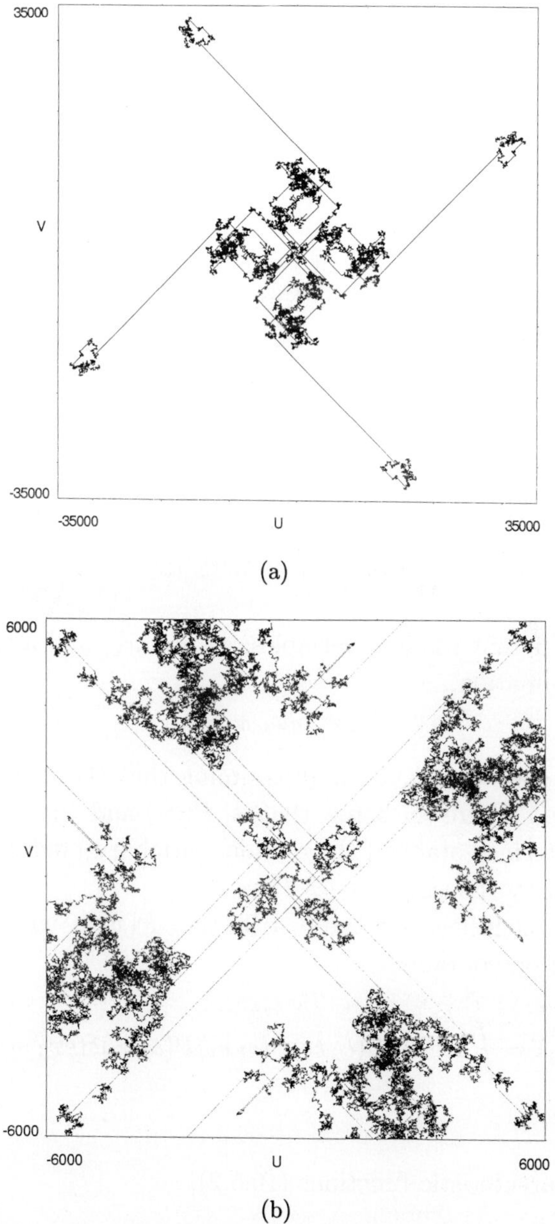


Fig. 10.4.2. A typical trajectory with anomalously long flights for the web-map with (a) $K = 6.349972$ and (b) a magnification of its central part.

The persistence for a "fast" variable by memory results in a change from normal kinetics, as described by the FPK equation, to an anomalous one (which will be discussed in Chapter 11). Prior to such a description, some pre-requisite information related to Lévy processes is introduced.

## 10.5 Lévy Processes

The Lévy stable distribution of random variables was proposed as a possible alternative to the Large Numbers Law, where $P(x)$ is the normalised distribution density of a random variable $x$, that is,

$$\int P(x)dx = 1 \,. \tag{10.5.2}$$

The Fourier transform of $P(x)$ is the characteristic function

$$P(q) = \int e^{iqx} P(x)dx \,. \tag{10.5.2}$$

Given two different random variables, $x_1$ and $x_2$, and a third which is a linear combination,

$$cx_3 = c_1 x_1 + c_2 x_2 \,, \tag{10.5.3}$$

where $c, c_1, c_2$ are constants, and assuming that the distribution functions of $x_1$ and $x_2$ are the same, that is, $P(x_1)$ and $P(x_2)$, the distribution $P(x)$ is termed stable if the random variable $x_3$ defined by (10.5.3) is distributed by the same law as $P(x_3)$.

When the condition for the distribution $P(x)$ is written as stable, from (10.5.3) one obtains

$$P(cx_3)d(cx_3) = \int d(c_1x_1)d(c_2x_2)P(c_1x_1)P(c_2x_2)\delta(cx_3 - c_1x_1 - c_2x_2) \tag{10.5.4}$$

or for the characteristic functions (10.5.2),

$$P(cq) = P(c_1 q) \cdot P(c_2 q) \,. \tag{10.5.5}$$

Using the logarithm of (10.5.5), one arrives at the functional equation

$$\ln P(cq) = \ln P(c_1 q) + \ln P(c_2 q) \tag{10.5.6}$$

which has the solution

$$\ln P(cq) = c\frac{q}{|q|}|q|^\alpha\,, \quad c^\alpha = c_1^\alpha + c_2^\alpha\,. \tag{10.5.7}$$

It should be $\text{Re} \ln P(q) < 0$. For $\alpha = 2$, the solution (10.5.7) corresponds to the normal (Gaussian) distribution and for $\alpha = 1$, to the Cauchy distribution. The condition $P(x) > 0$ gives restrictions

$$0 < \alpha \leq 2\,. \tag{10.5.8}$$

Thus,

$$P(q) = \text{const.} \exp(-c|q|^\alpha)\,, \quad (c > 0)\,. \tag{10.5.9}$$

Using the Tauber theorem, the following asymptotics are obtained:

$$\begin{aligned} P(x) &= \int_{-\infty}^{\infty} e^{-iqx} P(q) dq \\ &= \text{const.} \int_{-\infty}^{\infty} dq \exp -iqx + c|q|^\alpha \sim \frac{\alpha c\Gamma(\alpha)}{\pi|x|^{\alpha+1}} \sin(\pi\alpha/2)\,, \\ & |x| \to \infty \end{aligned} \tag{10.5.10}$$

Expressions (10.5.9) and (10.5.10) are Lévy distributions.[6]

There are many different ways to derive the time-dependent Lévy distribution $P(x,t)$ from (10.5.9). One method is outlined in Chapter 11. Here one shall consider a heuristic derivation. In considering the transition probability density, $P(x't'|x''t'')$, and the Chapman-Kolmogorov equation,

$$\begin{aligned} P(x_0 t_0|x_N t_N) = \int dx_1 \cdots dx_{N-1} P(x_0 t_0|x_1 t_1) \\ \times\, P(x_1 t_1|x_2 t_2) \cdots P(x_{N-1} t_{N-1}|x_N t_N)\,, \end{aligned} \tag{10.5.11}$$

[6]For more information on the Lévy distributions and their applications, see [Le 37], [MS 84], [SZF 95] and [Ka 95].

the expression can be simplified by setting

$$t_{j+1} - t_j = \Delta t \, (\forall j) \,, \quad t_N - t_0 = N\Delta t \,. \tag{10.5.12}$$

It is also assumed that

$$P(x_j t_j | x_{j+1} t_{j+1}) = P(x_{j+1} - x_j, t_{j+1} - t_j) \tag{10.5.13}$$

and expression (10.5.11) is rewritten with the help of (10.5.12) and (10.5.13):

$$P(x_N - x_0, t_N - t_0) = \int dy_1 \cdots dy_N P(y_1, \Delta t) \cdots P(y_N, \Delta t) \tag{10.5.14}$$

where the variables $y_j = x_j - x_{j-1}$ are replaced. Expression (10.5.14) is the $N$-fold generalisation of (10.5.4) and one can immediately write its Fourier transform as being similar to (10.5.5):

$$P(cq) = P(c_1 q) \cdots P(c_N q) \tag{10.5.15}$$

with the condition

$$c^\alpha = \sum_{j=1}^{N} c_j^\alpha = N c_0 = \frac{(t_N - t_0)}{\Delta t} c_0 \,. \tag{10.5.16}$$

Putting $x_0 = 0$, $x_n = x$, $t_0 = 0$, $t_N = t$, the solutions of (10.5.15), which are similar to (10.5.7) and (10.5.9), can be written as

$$P(q, t) = \text{const.} \exp(-at|q|^\alpha) \,, \tag{10.5.17}$$

where

$$a = c_0 / \Delta t \,. \tag{10.5.18}$$

The only thing required at this point in time is a consideration of a limit $c_0 \to 0$, $\Delta t \to 0$, $a =$ const. which does not change the form of the solution (10.5.17). The idea employed here is known as an infinitely divisible process.[7]

[7]For more information on the infinitely divisible processes, see [Le 37], [Gn 63] and [Fe 66].

It follows from (10.5.17) through the inverse Fourier transform that

$$P(x,t) \sim \frac{\alpha a t \Gamma(\alpha)}{\pi |x|^{\alpha+1}} \sin(\pi\alpha/2)\,, \quad |x| \to \infty\,, \qquad (10.5.19)$$

which is analogous to (10.5.10). A general property of the Lévy distributions (10.5.10) and (10.5.19) is a power-like decrease as $|x| \to \infty$. Under the condition $\alpha < 2$, it yields

$$\langle x^2 \rangle = \infty\,, \qquad (10.5.20)$$

which is a crucial feature of the distributions. One is therefore able to understand why the Lévy distribution (or process) can be considered an alternative to the Gaussian one, which has a finite second moment. The specific behaviour of $P(x,t)$ in (10.5.19) does not define any characteristic space scale. This leads to the appearance of very long pieces of trajectories with almost regular behaviour, that is, flights.[8]

## 10.6 The Weierstrass Random Walk

The Weierstrass random walk is a good illustration of the Lévy distribution function.[9] When the random variable $x$ has only values $x \in \{x_n = \pm a^n\}$ with a constant $a$ and the probability of $x$ having a value $x_n$ is $b^{-n}$, the normalised probability of having a value $x$ is then

$$P(x) = \frac{b-1}{2b} \sum_{n=0}^{\infty} b^{-n} (\delta_{x,a^n} + \delta_{x,-a^n})\,, \quad (b > 1) \qquad (10.6.1)$$

where $\delta_{x,x_n}$ are the Kronecker symbols. The characteristic function,

[8]Flights, or Lévy flights, were discussed intensively in [Ma 82] in connection with Lévy distribution and fractal Brownian motion. The resultant theory is found in [MS 84] on the basis of random walk models. See also [SKZ 96] where the fractal Brownian motion is related to different physical models and to fractal time. For more information on the applications of Lévy flights to dynamical systems, see [ASZ 91], [CZ 91] and [ZEN 97].

[9]Weierstrass random walks were introduced and analysed in [HSM 81], [MHS 82] and [MS 84], and are examined below.

$$P(k) = \sum_x e^{ikx} P(x) = \frac{b-1}{b} \sum_{n=0}^{\infty} b^{-n} \cos(a^n k) , \tag{10.6.2}$$

has the form of the Weierstrass function.

In considering the Bernoulli scaling, expression (10.6.2) is similar to (6.5.6). Function $P(k)$ satisfies the functional equation,

$$P(k) = \frac{1}{b} P(ak) + \frac{b-1}{b} \cos k , \tag{10.6.3}$$

which is analogous to (6.5.7). To derive the asymptotics $x \to \infty$, one needs to consider $k \to 0$. If the expression (10.6.2) is not singular, then for small $k$

$$P(k) = 1 - \frac{1}{2} \langle x^2 \rangle k^2 + \cdots \tag{10.6.4}$$

and the Gaussian distributions arise for the variable $x$ with finite $\langle x^2 \rangle$. The Weierstrass function is known to be continuous everywhere but not differentiable anywhere else if

$$\mu_W = \ln a / \ln b > 1 . \tag{10.6.5}$$

In the same way, as in Section 6.5, $P(k)$ can be considered a sum of the normal and singular parts,

$$P(k) = P_n(k) + P_s(k) , \tag{10.6.6}$$

and derive for $k \to 0$,

$$\begin{aligned} P_n(k) &= 1 + \text{const.} \cdot k^2 + \cdots \\ P_s(k) &= |k|^{1/\mu_W} Q(k) + \cdots \end{aligned} , \tag{10.6.7}$$

where $Q(k)$ is periodic in $\ln k$ with period $\ln a$. Expressions (10.6.6) and (10.6.7) correspond to the Lévy distribution with $\alpha = 1/\mu_W < 1$.

A more general approach to such phenomena is presented in Chapter 11. Here, however, one should note that the main features of the process (10.6.1) are the scaling properties of the support (that is, the possible values of $x$ are scaled) and the corresponding probabilities. These features match the scaling properties of the chaotic dynamics near the islands' boundary, which were presented in Tables 3.5.1 and 3.5.2 in Section 3.5.

## Conclusions

1. In a very simplified form, the chaotic dynamics can be described by a kinetic equation of the FPK-type (normal diffusion equation). A more careful analysis leads to the anomalous diffusion process for which
$$\langle R^2 \rangle \sim t^{\mu}$$
with $\mu \neq 1$.
2. The existence of an accelerator mode in the standard map and web-map induces superdiffusion with $\mu > 1$.
3. Diffusion along the channels of webs with symmetry is, generally speaking, anomalous.
4. Anomalous diffusion in the phase space can be induced by the presence of singular zones with a hierarchical structure of islands.

*Chapter 11*

# FRACTIONAL KINETICS

## 11.1 Fractional Generalisation of the Fokker-Planck-Kolmogorov Equation (FFPK)

The previous chapters have demonstrated that a number of typical systems with chaotic dynamics do not obey the FPK equation and that some generalisation has to be included in the kinetic description of the fractal properties of motion. The need for such generalisation results from a new property of chaos which is described as the incompleteness of chaos due to the existence of islands in the phase space. In fact, the specific hierarchical structure of resonances provides the physical basis for replacing regular space-time with redefined fractional space-time. This section will consider the applications of the elements of fractional calculus. The requisite information on fractional integration and differentiation is found in Appendix 3.[1]

---

[1]The use of fractal derivatives is a compact way to write an integral representation of functions with power-like kernels. Another simplified way to consider fractional derivative is to make a Fourier-transform and obtain the fractional powers of the corresponding variables. For information on fractional calculus, the following books are recommended: [GS 64], [SKM 87] and [MR 93]. The key information is provided in Appendix 3. Typically, fractional derivatives surface in different problems where a renormalisation has been used. Mention is made only on kinetics appearing near the phase transition point [HH 77] and particle advection in convective flows [YPP 89]. Dynamical chaos generates numerous applications of fractional calculus to describe the processes taking place in the phase space and time simultaneously. In [SZK 93], the notion of "strange kinetics" has been used.

Adopting the same scheme in Section 10.1, the expansion

$$W(x,y;t+\Delta t) = W(x,y;t) + \Delta t^{\beta}\frac{\partial^{\beta}W(x,y;t)}{\partial t^{\beta}} \tag{11.1.1}$$

is assumed to exist for $\Delta t \to 0$ with $\beta$ being some constant, $0 < \beta \leq 1$ (see Appendix 3 for the fractional derivative definition). In fact, the existence of the limit is assumed to be

$$\lim_{\Delta t \to 0} \frac{W(x,y;t+\Delta t) - W(x,y;t)}{\Delta t^{\beta}} = \frac{\partial^{\beta}W(x,y;t)}{\partial t^{\beta}}. \tag{11.1.2}$$

Equation (11.1.1) replaces the regular definition (10.1.6) when $\beta = 1$. Rewriting (11.1.1) using the notation (10.1.9) yields

$$P(x,t+\Delta t) = P(x,t) + \Delta t^{\beta}\frac{\partial^{\beta}P(x,t)}{\partial t^{\beta}}. \tag{11.1.3}$$

The function $P(x,t)$ describes the large time-scale asymptotics. Analogous to (10.1.13), an expansion for the transfer probability, $W(x,y;\Delta t)$, is written for the infinitesimal $\Delta t$ which imposes the infinitesimal changes of the co-ordinate $|x-y|$:

$$W(x,y;\Delta t) = \delta(x-y) + A(y;\Delta t)\delta^{(\alpha)}(x-y) + B(y;\Delta t)\delta^{(\alpha_1)}(x-y). \tag{11.1.4}$$

The superscripts $(\alpha)$ and $(\alpha_1)$ refer to the fractional derivatives for order $\alpha$ and $\alpha_1$ with an appropriate $\alpha$ as the fractal space dimension characteristic, $0 < \alpha \leq 1$. The definition of the fractional derivative for $\delta$-function is given in Appendix 3. A specific feature of fractional expansions is that it is not uniquely defined and the real order of the third term on the right-hand side of (11.1.4) is not known. With some approximation, one can neglect the term with $B$. In particular, it is correct if $A =$ const. and $\alpha_1 > \alpha$. Hence the normalisation condition,

$$\int W(x,y;\Delta t)dy = 1, \tag{11.1.5}$$

is automatically satisfied.

In the approximation to be used, (11.1.4) is rewritten without the term with $B$ as

$$W(x,y;\Delta t) = \delta(x-y) + A(\Delta t)\delta^{(\alpha)}(x-y)\,. \tag{11.1.6}$$

It is then multiplied by $|x-y|^\alpha$ and integrated, which yields (see Appendix 3)

$$\Gamma(1+\alpha)A(\Delta t) = \int_{-\infty}^{\infty} dx|x-y|^\alpha W(x,y;\Delta t) \equiv \langle\langle |x-y|^\alpha \rangle\rangle\,. \tag{11.1.7}$$

Assuming that the limit

$$\mathcal{A} \equiv \Gamma(1+\alpha) \lim_{\Delta t\to 0} A(\Delta t)/\Delta t^\beta = \lim_{\Delta t\to 0} \frac{\langle\langle |x-y|^\alpha \rangle\rangle}{\Delta t^\beta}\,, \tag{11.1.8}$$

which is similar to (10.1.18), exists and coincides with (10.1.18) if $\beta = 1$ and $\alpha = 2$, the local properties of the transitional process are then represented in the existence of the limit (11.1.8) for appropriate values of $\alpha$ and $\beta$. The latter values are not supposed to have the trivial values of two and one, respectively.

Turning now to (11.1.1) and using Eq. (10.1.4) with the same notations of (10.1.8) and (10.1.9), one now has, (instead of (10.1.11),)

$$\frac{\partial^\beta P(x,t)}{\partial t^\beta} = \lim_{\Delta t\to 0} \frac{1}{(\Delta t)^\beta} \left\{ \int dy[W(x,y;\Delta t) - \delta(x-y)]P(y,t) \right\} \tag{11.1.9}$$

or, after substituting (11.1.6)–(11.1.8),

$$\frac{\partial^\beta P(x,t)}{\partial t^\beta} = \mathcal{A} \int dy\delta^{(\alpha)}(x-y)P(y,t)\,. \tag{11.1.10}$$

The right-hand side can be integrated by parts using the formula

$$\int \frac{d^\alpha g(x)}{dx^\alpha} f(x)dx = \int g(x) \frac{d^\alpha f(x)}{d(-x)^\alpha} dx \tag{11.1.11}$$

(see Appendix 3 for its derivation and definitions). From (11.1.10) and (11.1.11), one then obtains

$$\frac{\partial^\beta P(x,t)}{\partial t^\beta} = \mathcal{A}\frac{\partial^\alpha P(x,t)}{\partial(-x)^\alpha}\,. \tag{11.1.12}$$

This is a form of fractional kinetic equation known as FFPK. Nevertheless, it is not the final form. Therefore, one needs to further discuss the process underlying Eq. (11.1.12) in order to find its solutions with an appropriate boundary and initial conditions.

It was specified that $t > 0$ in (11.1.12) and $-\infty < x < \infty$. The latter means that flights can occur in both the directions of positive and negative $x$. Hence all the integrals from (11.1.9) to (11.1.11) are taken at interval $(-\infty, \infty)$. The definition of $A(\Delta t)$ and the corresponding $\mathcal{A}$ in (11.1.8) should be along both the positive and negative "phases" of flights with equal probabilities. To make it possible, definition (11.1.6) is changed to a symmetrical form,

$$W(x,y;\Delta t) = \delta(x-y) + A\frac{\partial^\alpha}{\partial|x|^\alpha}\delta(x-y)\,, \tag{11.1.13}$$

where the symmetrical derivative of order $\alpha$ has been introduced,

$$\frac{\partial^\alpha}{\partial|x|^\alpha} = \frac{c_\alpha}{2}\left(\frac{\partial^\alpha}{\partial x^\alpha} + \frac{\partial^\alpha}{\partial(-x)^\alpha}\right), \tag{11.1.14}$$

with a constant $c_\alpha$ that regulates a special case such as $\alpha = 1$. For example, if $c_\alpha = 1$, $\partial^\alpha/\partial|x|^\alpha$ disappears for $\alpha = 1$ and

$$c_\alpha = 1/\cos\frac{\pi\alpha}{2}\,, \tag{11.1.15}$$

(11.1.14) is finite for $\alpha = 1$. Equation (11.1.12) is then transformed into a symmetrical equation:

$$\frac{\partial^\beta P(x,t)}{\partial t^\beta} = \mathcal{A}\frac{\partial^\alpha P(x,t)}{\partial|x|^\alpha}\,, \quad (\beta \le 1,\, 0 < \alpha \le 2) \tag{11.1.16}$$

with the same definition (11.1.8) of $\mathcal{A}$.

The problem of the kinetic description of dynamics is reduced to the phenomenological Eq. (11.1.16) where $\alpha$ and $\beta$ should be defined from the dynamics. This is carried out in Section 11.3.

Insofar as there is a refusal to use a regular expansion of series and consider fractional derivatives like in (11.1.3) and (11.1.6), more specific definitions and explanations of what kind of problem is to be considered are needed. It can be said that some level of analytical universality is lost and that one should determine all the limit procedures. For example, for Eqs. (11.1.6) and (11.1.3), it can be seen from Appendix 3 that the fractional derivatives are defined for the signed-fixed domain of a variable. This means that instead of using a derivative, say, $\partial^\alpha/\partial x^\alpha$, one can introduce a combination of $c^+\partial^\alpha/\partial x^\alpha + c^-\partial^\alpha/\partial(-x)^\alpha$ with the freedom to manipulate via $c^+$ and $c^-$ as coefficients or even functions of $\alpha$. This was done in a symmetrical form from (11.1.13) to (11.1.15). The reason for introducing expansion (11.1.13) can be explained in the following way. Defining fractional derivatives with the help of Fourier transform, one derives

$$\begin{aligned} \frac{\partial^\alpha}{\partial x^\alpha} &\to (-iq)^\alpha = \exp(i\pi\alpha/2)\cdot|q|^\alpha \\ \frac{\partial^\alpha}{\partial(-x)^\alpha} &\to (-iq)^\alpha = \exp(-i\pi\alpha/2)\cdot|q|^\alpha \end{aligned} . \tag{11.1.17}$$

Introducing the Fourier transform of $P(x,t)$, one has

$$P(x,t) = \frac{1}{2\pi}\int_{-\infty}^{\infty} dq\, e^{-iqx} P(q,t) . \tag{11.1.18}$$

Equation (11.1.16), together with definitions (11.1.14) of the symmetrical derivative and (11.1.15) of $c_\alpha$, gives

$$\frac{\partial^\beta P(q,t)}{\partial t^\beta} = \mathcal{A}|q|^\alpha P(q,t) , \tag{11.1.19}$$

which for $\beta = 1$ has the solution

$$P(q,t) = \text{const.}\cdot\exp(\mathcal{A}t|q|^\alpha) . \tag{11.1.20}$$

This is the Lévy process (10.5.17), which means that the choice of the derivative (11.1.14) is dictated by the necessity to satisfy the condition of coincidence of the FFPK Eq. (11.1.16) with the Lévy process for $\beta = 1$. However, this is not necessary in a general case and one must permit a reasonable amount of freedom in the use of fractional derivatives in expansion (11.1.6) where both signs of the variable are admissible.[2]

## 11.2 Evolution of Moments

Without the aid of any additional information, the FFPK Eq. (11.1.15) enables one to find the time-dependence of moments for the distribution $P(x,t)$. By definition, a moment of order $\eta$ is

$$\langle |x|^\eta \rangle = \int_{-\infty}^{\infty} dx |x|^\eta P(x,t) . \qquad (11.2.1)$$

After multiplying (11.1.15) by $|x|^\alpha$ and integrating it with respect to $x$ and $t$ as well as applying (11.1.11), one derives

$$\langle |x|^\alpha \rangle = \text{const} t^\beta . \qquad (11.2.2)$$

In the same way, one can find

$$\frac{\partial^\beta}{\partial t^\beta} \langle |x|^{2\alpha} \rangle = \text{const} \langle |x|^\alpha \rangle = \text{const} t^\beta \qquad (11.2.3)$$

or

$$\langle |x|^{2\alpha} \rangle = \text{const} t^{2\beta} . \qquad (11.2.4)$$

Similarly, the iteration of (11.2.2) to (11.2.4) yields

$$\langle |x|^{m\alpha} \rangle = \text{const} t^{m\beta} \qquad (11.2.5)$$

with integer $m$, or approximately

$$\langle x^{2m} \rangle \sim t^{m\mu} \qquad (11.2.6)$$

[2]FFPK-equation was introduced in [Za 92] and developed in [Za 94] and [Za 95] using the renormalisation group method. The solutions to the FFPK can be found in [SZ 97].

with

$$\mu = 2\beta/\alpha\,. \tag{11.2.7}$$

Speaking about $m > 1$, one ought to remember that the power-like tails of the distribution function can lead to infinite moments $\langle |x|^{m\alpha} \rangle$ starting from an appropriate $m$. At the same time, if the initial distribution function $P(x, 0)$ is localised, that is, $P(x, 0) = 0$ for $|x| > x_0$, then for any $t$ all moments $\langle |x|^{m\alpha} \rangle$ are finite because of the finite propagation velocity of particles and the boundary $x_0$ which simply moves with time. In particular, expression

$$\langle x^2 \rangle \sim t^{\mu} \tag{11.2.8}$$

defines the exponent $\mu$ which is equivalent to one for normal diffusion. At the same time, one should be careful to remember that (11.2.8) is an asymptotic expression and it is only valid for a limited period of time.[3]

## 11.3 Method of the Renormalisation Group for Kinetics (RGK)

A general space-time renormalisation procedure can be introduced for the kinetic equation of either (11.1.12) or (11.1.15) type without specifying the variable $x$. In a sense, the proposed procedure has an analog in condensed matter physics for kinetics near a phase transition point (see footnote [1]). Another example is the Weierstrass random walk which has been suggested as a model for Lévy processes (see Section 10.6). The main difference in the procedure proposed here is that the simultaneous space and time renormalisation is only qualitatively different from that in time (fractal time processes) or space.

Numerous examples have shown that a self-similarity of chaotic trajectories exists in those parts of trajectories which belong to the sticky motion near islands. In introducing the renormalisation group transform $\hat{R}_k$ for area $s_k$ and the last invariant curve period $T_k$ of the $k$-th generation island,

[3]For more information on moments and time restrictions, see [SZ 97].

$$\hat{R}_k \colon s_{k+1} = \lambda_s s_k \,; \quad T_{k+1} = \lambda_t T_k \,, \tag{11.3.1}$$

where $\lambda_s, \lambda_t$ are the scaling constants to be discussed later (see also Section 3.5). More flexibility in the definition of (11.3.1) is needed if dynamical systems are considered. Specifically,

$$\lambda_s \leq 1 \,, \quad \lambda_t \geq 1 \tag{11.3.2}$$

with the cut-off values of $s_{\max}$ and $t_{\min}$, which can be attributed to the initial (largest) island area and the island's last invariant curve period, respectively:

$$s_{\max} = s_0 \,, \quad T_{\min} = T_0 \,. \tag{11.3.3}$$

The inequalities in (11.3.2), together with the definitions of (11.3.3), indicate the direction of the $\hat{R}_k$-transformation process which proceeds to an arbitrarily small area, $s_n$, and then to an arbitrarily big value, $\Delta T_n$, of the last invariant curve period for a small island with area $s_n$.

The analysis of dynamical systems is different from the renormalisation approach to non-stationary critical phenomena in statistical physics. For that reason, one prefers not to use the conventional $(\omega, k)$ representation for kinetics. The main differences lie in the way the renormalisation of time and the replacement of the configurational space by the phase space is performed. The same difference would arise if the techniques presented here are compared to the Weierstrass random walk or with the anomalous transport phenomena in disordered systems.

As mentioned above, the self-similarity of the wandering process takes place as a result of the entanglement of trajectories in the singular zone of islands-around-islands. More self-similar relations can be introduced to reflect the phenomenon of entangling:

$$\begin{aligned} \sigma_{n+1} &= \lambda_\sigma \sigma_n \,, \quad \lambda_\sigma \leq 1 \\ \ell_{n+1} &= \lambda_\ell \ell_n \,, \quad \lambda_\ell \geq 1 \\ q_{n+1} &= \lambda_q q_n \,, \quad \lambda_q \geq 1 \end{aligned} \tag{11.3.4}$$

where $\sigma$ is the Lyapunov exponent, $\ell$ is the length of a flight, $q$ is the number of islands from the same generation, and $\lambda_\sigma, \lambda_\ell, \lambda_q$ are the corresponding scaling parameters.

The meaning of (11.3.4) can be gathered from Fig. 6.4.1 which presents a boundary layer near an island. The boundary layer involves a boundary islands chain (BIC) with $q$ islands ($q = 6$ in Fig. 6.4.1). Dividing the annulus by segments with an island in each of them, and taking into consideration the magnification of each segment, one obtains a similar annulus structure which represents BIC of the next, say, $(n+1)$-th generation if the initial BIC was of the $n$-th generation. All the sets $\{\lambda_s, \sigma_n, T_n, \ell_n, q_n\}$ belong to the domain $\text{BIC}_n \backslash \text{BIC}_{n+1}$.

As mentioned in Section 3.5, not all the scaling parameters are independent. In a typical situation, the estimation

$$\lambda_\ell = 1/\lambda_s^{1/2} \tag{11.3.5}$$

can be used, because of the area-preserving property of a cylindric layer along the flight with a cross-section of area $S$. Moreover, one can put

$$\lambda_T = 1/\lambda_\sigma \tag{11.3.6}$$

and for some simplified situation,

$$\lambda_T = \lambda_q\,, \tag{11.3.7}$$

as demonstrated in the tables in Section 3.5.

For all the scaling parameters of $\lambda_s, \lambda_\sigma, \lambda_T, \lambda_\ell, \lambda_q$, their corresponding scaling exponents are introduced:

$$\begin{gathered} \alpha_s = \alpha_0/|\ln \lambda_s|\,, \quad \alpha_\sigma = \alpha_0/|\ln \lambda_\sigma|\,, \quad \alpha_T = \alpha_0/\ln \lambda_T\,, \\ \alpha_\ell = \alpha_0/\ln \lambda_\ell\,, \quad \alpha_q = \alpha_0/\ln \lambda_q\,. \end{gathered} \tag{11.3.8}$$

The constant $\alpha_0$ is introduced to characterise the space in which the fractal set of islands is embedded. All the exponents can be connected to the generalised fractal dimensions of Section 5.3.

It was shown in Section 11.2 that the exponents $\alpha$ and $\beta$ of the space-time derivatives define the anomalous transport exponent $\mu$. Since the anomalous transport is a result of the critical hierarchical structure in the phase space and critical dynamics in the BIC zone (singular zone),

$\alpha$ and $\beta$ are therefore described as critical exponents. They will be defined in this section through the application of the RGK transform (11.3.1).

One now turns to a consideration of an infinitesimal time-deviation of the distribution function

$$\delta_t P(\ell, t) \equiv W(\ell, \ell_0, t + \Delta t) - W(\ell, \ell_0; t) , \tag{11.3.9}$$

where the notation (10.1.9) is used and the co-ordinate $x$ is replaced by the length, $\ell$, of a flight. This deviation can be expressed in accordance with the basic Eq. (10.1.4) through all the available paths and their probabilities. Thus,

$$\delta_t P(\ell, t) = \sum_{\Delta\ell} \overline{\{P(\ell + \Delta\ell, t) - P(\ell, t)\}} \equiv \delta_\ell P(\ell, t) , \tag{11.3.10}$$

where the sum over $\Delta\ell$ refers to the summation over all such paths within the infinitesimal interval $\Delta\ell$, which gives the infinitesimal evolutional change $\delta_t P$, and the bar means averaging over the paths. In particular, the direction of both flights are included. One can take the conventional averaging over the phases as an example. Equation (11.3.10) can be rewritten as

$$(\Delta t)^\beta \frac{\partial^\beta P}{\partial t^\beta} = \sum_{\Delta\ell} (\overline{(\Delta\ell)^\alpha \mathcal{D}}) \frac{\partial^\alpha P}{\partial \ell^\alpha} , \tag{11.3.11}$$

where $\mathcal{D}$ is some expansion constant.

Applying the RGK transform (11.3.1) in Eq. (11.3.10), one has

$$\hat{R}_K^n(\delta_t P) = \hat{R}_K^n(\delta_\ell P) . \tag{11.3.12}$$

Using (11.3.11), the following is derived:

$$[\hat{R}_K^n(\Delta t)^\beta] \frac{\partial^\beta P}{\partial t^\beta} = \sum_{\Delta t} \overline{[\hat{R}_K^n(\Delta\ell)^\alpha] \mathcal{D}} \frac{\partial^\alpha P}{\partial \ell^\alpha} . \tag{11.3.13}$$

Bearing in mind that the characteristic time scaling for the wandering of a particle in the singular zone of islands-around-islands (BIC of all generations) is simply the escape time scaling, one can put $\lambda_t = \lambda_T$.

This is in accordance with (11.3.6) because the Lyapunov exponent scaling defines the escape time scaling. For the space (length of a flight) transform, one can use the scaling parameter $\lambda_\ell$ after taking into account the relation (11.3.5). The RGK transform is now

$$\hat{R}_K\colon \Delta\ell_{k+1} = \lambda_\ell \Delta\ell_k\,; \quad \Delta t_{k+1} = \lambda_T \Delta t_k\,. \tag{11.3.14}$$

After substituting (11.3.14) in (11.3.13), one has

$$(\Delta t)^\beta \frac{\partial^\beta P}{\partial t^\beta} = \left(\frac{\lambda_\ell^\alpha}{\lambda_t^\beta}\right)^n \sum_{\Delta\ell} \overline{(\Delta\ell)^\alpha \mathcal{D} \frac{\partial^\alpha P}{\partial \ell^\alpha}} \tag{11.3.15}$$

or, in a more convenient form,

$$\frac{\partial^\beta P}{\partial t^\beta} = \left(\frac{\lambda_\ell^\alpha}{\lambda_t^\beta}\right)^n \sum_{\Delta\ell} \overline{\frac{(\Delta\ell)^\alpha}{(\Delta t)^\beta} \mathcal{D} \frac{\partial^\alpha P}{\partial \ell^\alpha}}\,. \tag{11.3.16}$$

The formula (11.3.16) is the renormalisation group equation for anomalous kinetics.

Beginning with $\ell_{\min} = \ell_0 = 1/S_0^2$ and $t_{\min} = T_0$ [see (11.3.3)], one has $(\lambda_\ell^\alpha/\lambda_t^\beta)^n \to 0$ or $\to \infty$ unless

$$\lambda_\ell^\alpha = \lambda_T^\beta\,. \tag{11.3.17}$$

Equation (11.3.17) defines a non-trivial "fixed point" for (11.3.16). It yields

$$\mu_0 \equiv \beta/\alpha = \ln\lambda_\ell / \ln\lambda_T \tag{11.3.18}$$

and the "fixed-point-kinetic-equation"

$$\frac{\partial^\beta P}{\partial t^\beta} = \sum_{\Delta\ell} \overline{\frac{(\Delta\ell)^\alpha}{(\Delta t)^\beta}} \mathcal{D} \frac{\partial^\alpha P}{\partial \ell^\alpha}\,, \tag{11.3.19}$$

which coincides with (11.1.16) if one identifies $\partial^\alpha/\partial|x|^\alpha$ with $\partial^\alpha/\partial\ell^\alpha$ and puts

$$\mathcal{A} = \sum_{\Delta\ell} \overline{\frac{(\Delta\ell)^\alpha}{(\Delta t)^\beta}} \mathcal{D}\,. \tag{11.3.20}$$

There are a few important consequences of (11.3.19) and (11.3.20). The kinetic equation in the form of

$$\frac{\partial^\beta P}{\partial t^\beta} = \mathcal{A}\frac{\partial^\alpha P}{\partial \ell^\alpha} \tag{11.3.21}$$

[see also (11.1.16)] now appears as a fixed point of the RGK. The condition for the existence of the fixed point leads to a connection between the space-time fractional exponents, $\alpha$ and $\beta$, and the space-time scales, $\lambda_\ell$ and $\lambda_T$. Expression (11.3.18) defines only the ratio $\beta/\alpha$. Therefore, using (11.3.8), one can write

$$\alpha = \alpha_\ell \cdot \alpha_0 = \alpha_0/\ln\lambda_\ell\,, \quad \beta = \alpha_T \cdot \alpha_0 = \alpha_0/\ln\lambda_T \tag{11.3.22}$$

with some constant $\alpha_0$ chosen in such a way that in the absence of the fractal support (due to a singular zone in the phase space), the trivial values of $\alpha = 2$ and $\beta = 1$ can be reached.

Equation (11.3.21) leads straight to the moment equation

$$\langle \ell^\alpha \rangle \sim \mathcal{A}t^\beta\,, \quad (t \to \infty)\,, \tag{11.3.23}$$

which gives the asymptotic law ($t$ is finite, $\ell < \ell_{\max}$):

$$\langle \ell^2 \rangle \sim t^{2\beta/\alpha} = t^\mu\,. \tag{11.3.24}$$

Comparing (11.3.24) to (11.2.8) and (1.3.18) to $\ell^2 = x^2$, one derives

$$\mu = 2\mu_0 = 2\beta/\alpha = 2\ln\lambda_\ell/\ln\lambda_T = |\ln\lambda_s|/\ln\lambda_T\,. \tag{11.3.25}$$

This is the final formula that expresses the transport exponent $\mu$ through the scaling properties of the singular zone of islands-around-islands. In fact, the scaling parameters $\lambda_s$ and $\lambda_T$ can be derived directly from the equations of motion or can be found from experiment or simulation.[4]

[4]The RGK method is described in [Za 94-2], [Za 94-3] and [ZEN 97].

## Conclusions

1. Fractional kinetics is designated to describe a random walk with scaling properties in space and time simultaneously. The corresponding FFPK equation for some situations is

$$\frac{\partial^\beta P(x,t)}{\partial t^\beta} = \mathcal{A}\frac{\partial^\alpha P(x,t)}{\partial |x|^\alpha}$$

with the fractional values of $\alpha$ and $\beta$.

2. The evolution of moments can be obtained from

$$\langle |x|^\alpha \rangle \sim t^\beta$$

or

$$\langle |x|^2 \rangle \sim t^\mu$$

with the transport exponent

$$\mu = 2\beta/\alpha\,.$$

3. The FFPK equation can be applied to chaotic dynamics when a hierarchical structure of islands is present. The method of the Renormalisation Group yields

$$\mu = |\ln \lambda_S| / \ln \lambda_T$$

with the area scaling exponent being $\lambda_S$ and the time scaling exponent, $\lambda_T$.

*Appendix 1*

# THE NONLINEAR PENDULUM

In considering the many problems associated with oscillations, the model of the nonlinear pendulum is a good approximation. Its Hamiltonian is

$$H_0 = \frac{1}{2}\dot{x}^2 - \omega_0^2 \cos x\,, \tag{A.1.1}$$

where a unit mass is assumed, that is, $p = \dot{x}$ and $\omega_0$ is the frequency of weak oscillations. The equation of motion for a nonlinear pendulum is written as

$$\ddot{x} + \omega_0^2 \sin x = 0 \tag{A.1.2}$$

and its phase portrait is shown in Fig. 1.4.2. The singularities are of the elliptic type ($\dot{x} = 0$, $x = 2\pi n$) and saddles [$\dot{x} = 0$, $x = 2\pi(n+1)$] and $n = 0, \pm 1, \ldots$. When $H_0 < \omega_0^2$, the trajectories correspond to the pendulum's oscillations (finite motion). However, when $H_0 > \omega_0^2$, they correspond to the pendulum's rotation (infinite motion because the phase $x$ grows infinitely). Trajectories with $H_0 = \omega_0^2$ are separatrices. The solution on the separatrix can be obtained if

$$H = H_s = \omega_0^2 \tag{A.1.3}$$

is substituted in Eq. (A.1.1). This yields the following equation:

$$\dot{x} = \pm 2\omega_0 \cos(x/2)\,. \tag{A.1.4}$$

For the initial condition $t = 0$, $x = 0$, the solution takes the form of

$$x = 4 \arctan \exp(\pm\omega_0 t) - \pi\,, \tag{A.1.5}$$

which is called a kink. Two whiskers of a separatrix (one entering the saddle and another leaving it) correspond to different signs of $t$. With the help of (A.1.4), the following is obtained from (A.1.5):

$$v = \dot{x} = \pm \frac{2\omega_0}{\cosh \omega_0 t} . \tag{A.1.6}$$

This solution is called a soliton.

In finding a solution for Eq. (A.1.1), a new parameter,

$$\kappa^2 = \frac{\omega_0^2 + H}{2\omega_0^2} = \frac{H + H_s}{2H_s} , \tag{A.1.7}$$

is introduced which defines the dimensionless energy with another initial value. The action can be obtained using formula (1.4.7):

$$I = I(H) = \frac{8}{\pi}\omega_0 \begin{cases} E(\pi/2;\kappa) - (1-\kappa^2)F(\pi/2;\kappa) , & (\kappa \le 1) \\ \kappa E(\pi/2; 1/\kappa) , & (\kappa \ge 1) \end{cases} , \tag{A.1.8}$$

where $F(\pi/2;\kappa)$ and $E(\pi/2;\kappa)$ are the complete elliptic integrals of the first and second kind, respectively. From the definition of the frequency of oscillations provided in (1.4.9) and (A.1.8), it follows that

$$\omega(H) = \frac{dH(I)}{dI} = \left[\frac{dI(H)}{dH}\right]^{-1} = \frac{\pi}{2}\omega_0 \begin{cases} 1/F(\pi/2;\kappa) , & (\kappa \le 1) \\ \kappa/F(\pi/2;1/\kappa) , & (\kappa \ge 1) \end{cases} . \tag{A.1.9}$$

The solution of the equation of motion (1.4.2) for velocity $\dot{x}$ is

$$\dot{x} = 2\kappa\omega_0 \begin{cases} \mathrm{cn}(t;\kappa) , & (\kappa \le 1) \\ \mathrm{dn}(t;\kappa) , & (\kappa \ge 1) \end{cases} , \tag{A.1.10}$$

where cn and dn are the Jacobian elliptic functions. When $\kappa = 1$, the expressions in (A.1.10) take the form of (A.1.6). An expansion of (A.1.10) into a Fourier series yields

$$\dot{x} = 8\omega \sum_{n=1}^{\infty} \frac{a^{n-1/2}}{1+a^{2n-1}} \cos[(2n-1)\omega t]\,, \quad (\kappa \le 1)$$
$$\dot{x} = 8\omega \left\{ \sum_{n=1}^{\infty} \frac{a^{n}}{1+a^{2n}} \cos n\omega t + \frac{1}{4} \right\}, \quad (\kappa \ge 1)\,, \tag{A.1.11}$$

where

$$\begin{aligned} a &= \exp(-\pi F'/F) \\ \omega &= \omega(H) \\ F &= F(\pi/2; \bar{\kappa}) \\ F' &= F[\pi/2; (1-\bar{\kappa}^2)^{1/2}] \\ \bar{\kappa} &= \begin{cases} \kappa & (\kappa \le 1) \\ 1/\kappa & (\kappa \ge 1)\,. \end{cases} \end{aligned} \tag{A.1.12}$$

Using the asymptotic expansions of the elliptic integrals, one derives

$$F(\pi/2; \kappa) \sim \begin{cases} \pi/2\,, & (\kappa \ll 1) \\ \dfrac{1}{2} \ln \dfrac{32 H_s}{H_s - H}\,, & (1-\kappa^2 \ll 1) \end{cases}. \tag{A.1.13}$$

The asymptotics for the parameter of expansion $a$ can be obtained from (A.1.12) and (A.1.13):

$$a \sim \begin{cases} \kappa^2/32\,, & (\kappa \ll 1) \\ \exp(-\pi/N_0)\,, & (1-\kappa^2 \ll 1) \end{cases}, \tag{A.1.14}$$

where a new parameter, $N_0$, is introduced:

$$N_0 = \frac{\omega_0}{\omega(H_o)} = \frac{2}{\pi} F(\pi/2; \kappa)\,, \tag{A.1.15}$$

which determines the ratio of the frequency $\omega_0$ of the pendulum's small oscillations to its frequency at a given energy $H_0$. According to (A.1.13), its asymptotic expansions are

$$N_0 \sim \begin{cases} 1\,, & (\kappa \ll 1) \\ \dfrac{1}{\pi} \ln \dfrac{32 H_s}{H_s - H_0}\,, & (1 - \kappa^2 \ll 1) \end{cases} . \tag{A.1.16}$$

From (A.1.14), it follows that the number $N_0$ determines the effective number of harmonics in the Fourier spectrum. If $n > N_0$, the amplitudes decrease exponentially. Otherwise, if $n < N_0$, all the coefficients in the Fourier transform would be of the same order of magnitude.

The foregoing has provided a full description of the character of the pendulum's oscillations. The case of $\kappa \ll 1$ corresponds to small amplitudes of oscillations when $H_0$ is close to $-H_s$ (that is, oscillations occur near the bottom of the potential well (see Fig. 1.4.1)). In accordance with (A.1.14) and (A.1.16), $\omega(H_0) \approx N_0 \sim 1$ and the amplitude $a$ is small. Consequently, in the equations in (A.1.11), one needs only retain the first term (which corresponds to linear oscillations),

$$v = \dot{x} \approx (\delta H)^{1/2} \cos \omega_0 t\,,$$

where

$$\delta H = H_s + H_0 = \omega_0^2 - |H_0|\,.$$

In the case of $\kappa^2 \to 1$, that is, $H_0 \to H_s$, the frequency $\omega(H_0) \to 0$ and the period of oscillation diverges logarithmically in the vicinity of the separatrix. As a function of time, the pendulum's velocity $\dot{x}$ consists of an approximate periodic sequence of soliton-like pulses (Fig. 1.4.3). The interval between two adjacent bumps in the same phase hardly differs from the period of oscillation $2\pi/\omega(H_0)$, and the width of each pulse is close to $2\pi/\omega_0$. Thus, the number $N_0$ also determines the ratio of the distance between the soliton-like pulses to their width when the oscillations in the neighbourhood of the separatrix are being considered.

The properties of the oscillations described above are true not only when one approaches the separatrix from below, that is, $(H_0 \to \omega_0^2 - 0)$, but also for $(H_0 \to \omega_0^2 + 0)$ when it is approached from the side of infinite trajectories.

*Appendix 2*

# SOLUTION TO THE RENORMALISATION TRANSFORM EQUATION

The definition of (6.5.6) for the waiting function (exit time distribution) $\psi(t)$ is

$$\psi(t) = \frac{1-a}{a}\sum_{j=1}^{\infty}(ab)^j \exp(-b^j t)\,. \tag{A.2.1}$$

In deriving its asymptotic representation, one follows [MS 84].

The Laplace transform

$$\psi(u) = \int_0^\infty e^{-ut}\psi(t)dt \tag{A.2.2}$$

transforms (A.2.1) into

$$\psi(u) = \frac{1-a}{a}\sum_{j=1}^{\infty}\frac{(ab)^j}{u+b^j} = \frac{1-a}{2\pi i a}\sum_{j=1}^{\infty}(ab)^j\int_{\sigma-i\infty}^{\sigma+i\infty}\frac{\pi}{\sin\pi\xi}u^{-\xi}b^{j(\xi-1)}d\xi\,, \tag{A.2.3}$$

where the Mellin transform,

$$f(\xi) \equiv [Mf(t)] = \int_0^\infty f(t)t^{\xi-1}dt\,, \tag{A.2.4}$$

has been applied to $1/(u+b^j)$. Replacing the summation and integration in (A.2.3) yields

$$\psi(u) = \frac{1-a}{a}\frac{1}{2\pi i}\int_{\sigma-i\infty}^{\sigma+i\infty} d\xi \frac{\pi u^{-\xi}}{\sin\pi\xi}\cdot\frac{1}{1-ab^{\xi}}\,. \tag{A.2.5}$$

The simple poles of the under-integral expression are in points

$$\xi = 0, \pm 1, \ldots, \quad \xi = -\ln a/\ln b \pm 2\pi i n/\ln b\,, \quad (n = 0, 1, \ldots)\,. \tag{A.2.6}$$

Taking the residues from all the poles in (A.2.6), one derives

$$\begin{aligned} \psi(u) &= 1 + u^{\ln a/\ln b} Q(u) + \frac{1-a}{a}\sum_{j=1}^{\infty}\frac{(-1)^n a u^n}{b^n - a} \\ Q(u) &= \frac{1-a}{n\ln b}\sum_{n=-\infty}^{\infty}\frac{\pi a b^{\delta_n}}{\sin\pi\delta}\exp(-2\pi i n \ln u/\ln b) \end{aligned} \tag{A.2.7}$$

with

$$\delta_n = -\ln a/\ln b + 2\pi i n/\ln b\,. \tag{A.2.8}$$

The singular behaviour of $\psi(u)$ is caused only by the second term in (A.2.7),

$$\psi(u) \sim u^{\beta''} Q(u)\,, \quad (u \to 0)\,, \tag{A.2.9}$$

which defines the asymptotics

$$\psi(t) \sim t^{-1-\beta''}\,, \quad (t \to \infty) \tag{A.2.10}$$

with

$$\beta'' = \ln a/\ln b\,. \tag{A.2.11}$$

The idea of deriving the exponent $\beta''$ for the function $\psi(t)$ can be used on other equations possessing the same renormalisation property as Eq. (A.2.1).

*Appendix 3*

# FRACTIONAL INTEGRO-DIFFERENTIATION

Here, some useful formulas of fractional calculus used to derive the FFPK equation are given. They are also found in [GS 64], [SKM 87] and [MR 93]. The idea of a fractional integral can be formulated by using the Cauchy formula,

$$\begin{aligned} g_n(x) &= \int_a^x \int_x^{x_{n-1}} \cdots \int_a^{x_1} g(\xi) d\xi dx_1 \cdots dx_{n-1} \\ &= \frac{1}{(n-1)!} \int_a^x g(\xi)(x-\xi)^{n-1} d\xi \,, \end{aligned} \tag{A.3.1}$$

and in its shortened form,

$$g_n(x) = g(x) \star \frac{x_+^{n-1}}{\Gamma(n)} \,, \tag{A.3.2}$$

where the sign $\star$ means convolution and $x_+$ is a generalised function defined on the semi-axis $x > 0$. Hence fractional integration is a generalisation of the definition of (A.3.2) for an arbitrary value of $\alpha$ instead of the integer $n$:

$$g_\alpha(x) = g(x) \star x_+^{\alpha-1}/\Gamma(\alpha) \,. \tag{A.3.3}$$

Expression (A.3.3) defines the fractional integration for order $\alpha$ if $\alpha > 0$. One can further consider

$$g_{-\alpha}(x) = g(x) \star x_+^{-\alpha-1}/\Gamma(-\alpha) \tag{A.3.4}$$

as the definition for a fractional derivative of the order $\alpha$.

Thus, one arrives at the so-called Riemann-Liouville definition of fractional integro-differentiation. More general (and less convenient) notations are used for this purpose:

(i) The left-hand side integro-differentiation is

$${}_a\mathcal{D}_+^\alpha g(x) \equiv g(x) \star \frac{x_+^{-(\alpha+1)}}{\Gamma(-\alpha)} = \int_a^x dx' g(x') \frac{(x-x')^{-(\alpha+1)}}{\Gamma(-\alpha)} . \tag{A.3.5}$$

(ii) The right-hand side integro-differentiation is

$${}_b\mathcal{D}_-^\alpha g(x) \equiv -g(x) \star \frac{x_-^{-(\alpha+1)}}{\Gamma(-\alpha)} = \int_x^b dx' g(x') \frac{(x-x')^{-(\alpha+1)}}{\Gamma(-\alpha)} , \tag{A.3.6}$$

where $x_-$ is a generalised function on the semi-axis $x < 0$.

Notably, the limits on the integration $(a, b)$ can be extended to $(-\infty, \infty)$. The definitions (A.3.5) and (A.3.6) can also be considered either on a full axis or on a semi-axis if corresponding integrals do exist.

The following formulas can be useful:

$${}_a\mathcal{D}_+^\alpha \cdot {}_a\mathcal{D}_+^\beta = {}_a\mathcal{D}_+^{\alpha+\beta} , \quad {}_b\mathcal{D}_-^\alpha \cdot {}_b\mathcal{D}_-^\beta = {}_b\mathcal{D}_-^{\alpha+\beta} . \tag{A.3.7}$$

In particular, one obtains the identity when $\beta = -\alpha$. Integro-differentiation can be applied to the $\delta$-function and its derivatives,

$${}_a\mathcal{D}_+^\alpha [\delta^{(k)}(x)] = \frac{x_+^{-k-\alpha-1}}{\Gamma(-k-\alpha)} , \tag{A.3.8}$$

or more generally, to

$${}_a\mathcal{D}_+^\alpha \left[ \frac{x_+^{\beta-1}}{\Gamma(\beta)} \right] = \frac{x_+^{-\beta-\alpha-1}}{\Gamma(\beta-\alpha)} . \tag{A.3.9}$$

When the scalar product of the two functions $f$ and $g$ is denoted as

$$[g(x) \cdot f(x)] \equiv \int_a^b dx\, g(x) f(x) \,, \tag{A.3.10}$$

the following formula is equivalent to an integration by parts:

$$[{}_a\mathcal{D}_+^\alpha g(x) \cdot f(x)] = [g(x) \cdot {}_b\mathcal{D}_-^\alpha f(x)] \,. \tag{A.3.11}$$

Finally, the following expression is needed:

$${}_a\mathcal{D}_+^\alpha x^p = \frac{\Gamma(p+1)}{\Gamma(p+1-\alpha)} x^{p-\alpha} \,. \tag{A.3.12}$$

For $p = \alpha$,

$${}_a\mathcal{D}_+^\alpha x^\alpha = \Gamma(1+\alpha) \,.$$

When $p = 1$, ${}_a\mathcal{D}_+^1 x = 1$, which is usually the case. In an unusual case when $p = 0$,

$${}_a\mathcal{D}_+^\alpha 1 = \frac{1}{\Gamma(1-\alpha)} x^{-\alpha} \,, \tag{A.3.13}$$

that is, a non-zero result is obtained for the fractional derivative of a constant unless $\alpha$ is of the integer $n \geq 1$.

Below is an analog of the Taylor expansion formula for $f(x)$ near a point $\xi$:

$$f(x) = f(\xi) + \sum_{j=0}^{n-1} \frac{(x-\xi)^{\alpha+j}}{\Gamma(\alpha+j+1)} {}_a\mathcal{D}_+^{\alpha+j} f(\xi) + R_n(x) \,. \tag{A.3.14}$$

The remainder term is $R_n(x)$. The estimations for $R_n(x)$ can be found in [SKM 87].

In Section 11.1, simplified notations are used for the fractional derivatives

$$\frac{\partial^\alpha g(x)}{\partial x^\alpha} \equiv {}_{-\infty}\mathcal{D}_+^\alpha g(x) \,, \quad \frac{\partial^\alpha g(x)}{\partial (-x)^\alpha} \equiv {}_\infty\mathcal{D}_-^\alpha g(x) \,. \tag{A.3.15}$$

More information on the application of the fractional calculus to FFPK equation can be found in [SZ 97].

# References


**A64** Arnold, V. I. (1964) *Dokl. Akad. Nauk SSSP* **156**, 9

**A88** ——— (1988) *Physica D* **33**, 21

**A89** ——— (1989) *Mathematical Methods of Classical Mechanics.* New York: Springer

**AB93** Abdullaev, S. S. (1993) *Chaos and Dynamics of Rays in Waveguide Media*, ed. G. M. Zaslavsky. New York: Harwood Acad. Publ.

**ACSZ90** Afanas'ev, V. V., Chernikov, A. A., Sagdeev, R. Z. & Zaslavsky, G. M. (1990) *Phys. Lett. A* **144**, 229

**Af97** Afraimovich, V. (1997) *Chaos* **7**, 12

**ASZ91** Afanas'ev, V. V., Sagdeev, R. Z. & Zaslavsky, G. M. (1991) *Chaos* **1**, 143

**Au79** Aubry, S. (1979) In *Solitions and Condensed Matter Physics*, ed. A. R. Bishop & T. Schneider, pp. 264. Heidelberg: Springer

**AZ91** Abdullaev, S. S. & Zaslavsky, G. M. (1991) *Sov. Phys. Uspekhi* **34**, 645

**B95** Boltzmann, L. (1895) *Nature* **51**, 413

**Ba75** Balescu, R. (1995) *Equilibrium and Non-Equilibrium Statistical Mechanics.* New York: John Wiley & Sons

**Ba95** —— (1995) *Phys. Rev. E* **51**, 4807

**Ba97** —— (1997) *Phys. Rev. E* **55**, 2465

**BBZT84** Belobrov, P. I., Beloshapkin, V. V., Zaslavsky, G. M. & Tretyakov, A. G. (1984) *Zhurn. Eksp. Teor. Fiz.* **87**, 310

**BG90** Bouchaud, J. P. & Georges, A. (1990) *Phys. Rep.* **195**, 127

**BKWZ97** Benkadda, S., Kassibrakis, S., White, R. & Zaslavsky, G. M. (1997) *Phys. Rev. E* **55**, 4909

**BL92** Bleher, P. M. (1992) *J. Stat. Phys.* **66**, 315

**BTZ94** Beloshapkin, V. V., Tretyakov, A. G. & Zaslavsky, G. M. (1994) *Comm. Pure and Appl. Math.* **47**, 39

**Bo72** Bowen, R. (1972) *Amer. J. Math.* **94**, 413

**BZ83** Beloshapkin, V. V. & Zaslavsky, G. M. (1983) *Phys. Lett. A* **97**, 121

**C59** Chirikov, B. V. (1959) *Atomnaya Energiya* **6**, 630 (in Russian)

**C79** —— (1979) *Phys. Rep.* **52**, 264

**C91** —— (1991) *Chaos, Solitons and Fractals* **1**, 79

**CET86** Cary, J. R., Escande, D. E. & Tennyson, J. L. (1986) *Phys. Rev. A* **34**, 4256

**CM81** Cary, J. R. & Meiss, J. D. (1981) *Phys. Rev. A* **24**, 2664

**CNPSZ87** Chernikov, A. A., Natenzon, M. Ya., Petrovichev, B. A., Sagdeev, R. Z. & Zaslavsky, G. M. (1987) *Phys. Lett. A* **122**, 39

**CNPSZ88** *ibid.* (1988) **129**, 377

**CR94** Chernikov, A. A. & Rogalsky, A. V. (1994) *Chaos* **4**, 35

**CS84** Chirikov, B. V. & Shepeliansky, D. L. (1984) *Physica D* **13**, 394

**CZ91** Chaikovsky, D. K. & Zaslavsky, G. M. (1991) *Chaos* **1**, 463

**D95** Dana, I. (1995) *Phys. Lett. A* **197**, 413

**DA95** Dana, I & Amit, M. (1995) *Phys. Rev. E* **51**, R2731

**DK96** Dana, I. & Kalitsky, T. (1996) *Phys. Rev. E* **53**, R2025

**DMP89** Dana, I., Murray, M. W. & Percival, I. C. (1989) *Phys. Rev. Lett.* **62**, 233

**EE59** Ehrenfest, P. & Ehrenfest, T. (1959) *The Conceptual Foundations of the Statistical Approach in Mechanics.* Translated by M. T. Moravesik. Ithaca, New York: Cornell University Press

**EE91** Elskens, Y. & Escande, D. F. (1991) *Nonlinearity* **4**, 615

**Fe66** Feller, V. (1966) *An Introduction to Probability Theory and Its Applications, Vol. 2.* New York: John Wiley & Sons

**FMIT77** Fukuyama, A., Momota, H., Itatani, R. & Takizuka, T. (1977) *Phys. Rev. Lett.* **38**, 701

**FP85** Frisch, U. & Parisi, G. (1985) In *Turbulence and Predictability of Geophysical Flows and Climate Dynamics*, ed. M. Ghill, R. Benzi & G. Parisi. Amsterdam: North-Holland

**FSZ67** Filonenko, N. N., Sagdeev, R. Z. & Zaslavsky, G. M. (1967) *Nuclear Fusion* **7**, 253

**GHP83** Grassberger, P., Procaccia, I. & Hentschel, M. (1983) *Lect. Notes Phys.* **179**, 212

**Gn63** Gnedenko, B. V. (1963) *The Theory of Probability.* New York: Chelsea Publ. Co.

**GNZ84** Geisel, T., Nierwetberg, J. & Zacherl, A. (1984) *Phys. Rev. Lett.* **54**, 616

**GP84** Grassberger, P. & Procaccia, I. (1984) *Physica D* **13**, 34

**Gr79** Green, J. M. (1979) *J. Math. Phys.* **20**, 1183

**GS94** Gelfand, I. M. & Shilov, G. E. (1964) *Generalised Functions, Vol. 1.* New York: Academic Press

**GS87** Grunbaum, B. & Shepard, G. C. (1987) *Tiling and Patterns.* New York: W. H. Freeman

**GZR87** Geisel, T., Zacherl, A. & Radons, G. (1987) *Phys. Rev. Lett.* **59**, 2503

**HH77** Hohenberg, P. C. & Halperin, B. I. (1977) *Rev. Mod. Phys.* **49**, 435

**HI96** Horton, W. & Ichikawa, Y.-H. (1996) *Chaos and Structures in Nonlinear Plasmas.* Singapore: World Scientific

**HJKPS86** Halsey, T. C., Jensen, M. H., Kadanoff, L. P., Procaccia, I. & Schraiman, B. I. (1986) *Phys. Rev. A* **33**, 1141

**HMS82** Hughes, B. D., Montroll, E. W. & Shlesinger, M. F. (1982) *J. Stat. Phys.* **28**, 111

**HP83** Hentschel, H. G. E. & Procaccia, I. (1985) *Physica D* **8**, 435

**HSM81** Hughes, B. D., Shlesinger, M. F. & Montroll, E. W. (1981) *Proc. Natl. Acad. Sci. USA* **78**, 3287

**IKH87** Ichikawa, Y.-H., Kamimura, T. & Hatori, T. (1987) *Physica D* **29**, 247

**Is92** Isichenko, M. B. (1992) *Rev. Mod. Phys.* **64**, 961

**JKLPS85** Jensen, M. H., Kadanoff, L. P., Libshaber, A., Procaccia, I. & Stavans, J. (1985) *Phys. Rev. Lett.* **55**, 439

**K38** Kolmogorov, A. N. (1938) *Uspekhi Matem. Nauk* **5**, 5

**Ka58** Kac, M. (1958) *Probability and Related Topics in Physical Sciences.* New York: Interscience

**Ka83** Karney, C. F. F. (1983) *Physica D* **8**, 360

**Ka95** Kahane, J.-P. (1995) In *Lévy Flights and Related Topics in Physics*, ed. M. F. Shlesinger, G. M. Zaslavsky & U. Frisch, pp. 99. New York: Springer

**Kl70** Klein, M. J. (1970) *Science* **169**, 361

**Kr50** Krylov, N. S. (1979) *Works on the Foundation of Statistical Physics, Moscow-Leningrad, Academy of Sciences 1950.* English translation. New York: Princeton University Press

**KSZ96** Klafter, J., Shlesinger, M. F. & Zumofen, G. (1996) *Physics Today* **2**, 33

**KZ97** Kuznetsov, L. I. & Zaslavsky, G. M. (1997) *Phys. Reports* **288**, 457

**L37** Landau, L. D. (1937) *Zhurn. Eksp. Teor. Fiz* **7**, 203

**La93** Lamb, J. S. W. (1993) *J. Phys. A* **26**, 2921

**Le37** Lévy, P. (1937) *Theorie de l'Addition des Variables Aletoires.* Paris: Gauthier-Villiers

**LiL93** Lichtenberg, J. & Lieberman, M. A. (1993) *Regular and Stochastic Motion.* Heidelberg: Springer

**LL76** Landau, L. D. & Lifshits, E. M. (1976) *Mechanics.* Oxford: Pergamon

**Lo91** Lowenstein, J. H. (1991) *Chaos* **1**, 473

**Lo92** —— (1992) *Chaos* **2**, 413

**Lo93** —— (1993) *Phys. Rev. E* **47**, R3811

**Lo94** —— (1994) *Chaos* (1994) **4**, 397

**Lo94** —— (1994) *Phys. Rev. E* **49**, 232

**LQ94** Lamb, J. S. W. & Quispel, G. R. W. (1994) *Physica D* **73**, 277

**LR90** Leff, H. S. & Rex, A. (1990) *Maxwell's Demon.* New Jersey: Princeton University Press

**LST89** Lazutkin, V. F., Schakhmanski, I. G. & Tabanov, M. B. (1989) *Physica D* **40**, 235

**LW89** Lichtenberg, A. J. & Wood, B. P. (1989) *Phys. Rev. A* **39**, 2153

**M86** Meiss, J. D. (1986) *Phys. Rev. A* **34**, 2375

**M92** —— (1992) *Rev. Mod. Phys.* **64**, 795

**M97** —— (1997) *Chaos* **7**, 139

**Ma82** Mandelbrot, B. (1982) *The Fractal Geometry of Nature.* New York: Freeman

**Mar69** Margulis, G. A. (1969) *Funkts. Anal. i Prilozh.* (Functional Analysis and Its Applications, in Russian) **3**, 80

**Mar70** —— (1970) *Funkts. Anal. i Prilozh.* **4**, 62

**Me63** Melnikov, V. K. (1963) *Dokl. Akad. Nauk SSSR* **148**, 1257

**Me96** —— (1996) In *Transport, Chaos and Plasma Physics 2*, ed. S. Benkadda, F. Doveil & Y. Elskens, pp. 142. Singapore: World Scientific

**MMP84** MacKay, R. S., Meiss, J. D. & Percival, I. C. (1984) *Physica* **13D**, 55

**MR93** Miller, K. S. & Ross, B. (1993) *An Introduction to the Fractional Differential Equations. New York: John Wiley & Sons*

**MS84** Montroll, E. W. & Shlesinger, M. F. (1984) In *Studies in Statistical Mechanics Vol. 11*, ed. J. Lebowitz & M. Montroll, pp. 1. Amsterdam: North-Holland

**N75** Neishtadt, A. I. (1975) *J. Appl. Math. Mech.* **39**, 594 (in Russian)

**NCC91** Neishtadt, A. I., Chaikovsky, D. K. & Chernikov, A. A. (1991) *Zhurn. Eksp. Teor. Fiz* **99**, 763

**Ne77** Nekhoroshev, N. N. (1977) *Uspekhi Matem. Nauk* **32**, 5

**NST97** Neishtadt, A. I., Sidorenko, V. V. & Treschev, D. V. (1997) *Chaos* **7**, 2

**P74** Penrose, R. (1974) *Bull. Inst. Math. App.* **10**, 266

**Pe88** Pesin, Ya. (1988) *Russ. Math. Surveys* **43**, 111

**HP83** Hentschel, H. G. E. & Procaccia, I. (1983) *Physica D* **8**, 435

**PR97** Pekarsky, S. & Rom-Kedar, V. (1997) forthcoming

**PV87** Paladin, G. & Vulpiani, A. (1987) *Phys. Rep.* **156**, 147

**PW97** Pesin, Ya. & Weiss, H. (1997) *Chaos* **7**, 89

**PY93** Parmenter, R. H. & Yu, Y. Y. (1991) *Chaos* **3**, 369

**RB81** Richens, P. J. & Berry, M. V. (1981) *Physica D* **2**, 495

**Re70** Rényi, A. (1970) *Probability Theory.* Amsterdam: North-Holland

**Ro94** Rom-Kedar, V. (1994) *Nonlinearity* **7**, 441

**Ro95** —— (1995) *Chaos* **5**, 385

**Ro97** —— (1997) *Chaos* **7**, 148

**RW80** Rechester, A. B. & White, R. B. (1980) *Phys. Rev. Lett.* **44**, 1586

**Sh88** Shlesinger, M. F. (1988) *Ann. Rev. Phys. Chem.* **39**, 269

**SKM87** Samko, S. G., Kilbas, A. A. & Marichev, O. I. (1987) *Fractional Integrals and Derivatives and Their Applications.* Translated by Harwood Academic Publishers. Minsk: Naukai Tekhnika

**SSB91** Scher, H., Shlesinger, M. F. & Bender, J. T. (1991) *Physics Today* **1**, 26

**SUZ89** Sagdeev, R. Z., Usikov, D. A. & Zaslavsky, G. M. (1989) *Nonlinear Physics.* New York: Harwood Academic Publishers

**Se95** Senechal, M. (1995) *Quasi-crystals and Geometry.* Cambridge: Cambridge University Press

**SZ97** Saichev, A. I. & Zaslavsky, G. M. (1997) Unpublished

**SZK93** Shlesinger, M. F., Zaslavsky, G. M. & Klafter, J. (1993) *Nature* **363**, 31

**SZU95** Shlesinger, M. F., Zaslavsky, G. M. & Frisch, U. (1995) *Lévy Flights and Related Topics in Physics.* Heidelberg: Springer

**Ta69** Taylor, J. B. (1969) *Culham Lab. Prog. Report CLM-PR-12*

**Tr96** Treschev, D. V. (1996) *Chaos* **6**, 6

**Tr97** —— (1997) *Physica D*, forthcoming

**We52** Weyl, H. (1952) *Symmetry.* New Jersey: Princeton University Press

**YP92** Yu, L. Y. & Parmenter, R. H. (1992) *Chaos* **2**, 581

**YPP89** Young, W., Pumir, A. & Pomeau, Y. (1989) *Phys. Fluids A* **1**, 462

**Za81** Zaslavsky, G. M. (1981) *Phys. Reports* **80**, 57

**Za85** —— (1985) *Chaos in Dynamic Systems.* New York: Harwood Academic Publishers

**Za92** —— (1992) In *Topological Aspects of the Dynamics of Fluids and Plasmas*, ed. H. K. Moffatt, G. M. Zaslavsky, P. Compte & M. Tabor, pp. 481. Dordrecht: Kluwer

**Za94** —— (1994) *Chaos* **4**, 589

**Za94-2** —— (1994) *Physica D* **76**, 110

**Za94-3** —— (1994) *Chaos* **4**, 25

**Za95** —— (1995) *Chaos* **5**, 653

**ZA95** Zaslavsky, G. M. & Abdullaev, S. S. (1995) *Phys. Rev. E* **51**, 3901

**Ze96** Zermelo, E. (1896) *Ann. Phys. Board* **57**, 485

**ZF68** Zaslavsky, G. M. & Filonenko, N. N. (1968) *Sov. Phys. JETP* **25**, 851

**ZE97** Zaslavsky, G. M. & Edelman, M. (1997) *Phys. Rev. E* **56**, 5310

**ZEN97** Zaslavsky, G. M., Edelman, M & Niyazov, B. (1997) *Chaos* **7**, 159

**ZN97** Zaslavsky, G. M. & Niyazov, B. (1997) *Physics Reports*, forthcoming

**ZSCC89** Zaslavsky, G. M., Sagdeev, R. Z., Chaikovsky, D. K. & Chernikov, A. A. (1989) *Sov. Phys. JETP* **68**, 995

**ZSUC91** Zaslavsky, G. M., Sagdeev, R. Z., Usikov, D. A. & Chernikov, A. A. (1991) *Weak Chaos and Quasi-regular Patterns.* Cambridge: University Press

**ZSW93** Zaslavsky, G. M., Stevens, D. & Weitzner, H. (1993) *Phys. Rev. E* **48**, 1683

**ZT91** Zaslavsky, G. M. & Tippett, M. K. (1991) *Phys. Rev. Lett.* **67**, 3251

**ZZNSUC89** Zaslavsky, G. M., Zakharov, M. Yu., Neishtadt, A. I., Sagdeev, R. Z., Usikov, D. A. & Chernikov, A. A. (1989) *Sov. Phys. JETP* **69**, 1563

**ZZSUC86** Zaslavsky, G. M., Zakharov, M. Yu., Sagdeev, R. Z., Usikov, D. A. & Chernikov, A. A. (1986) *Sov. Phys. JETP* **64**, 294

# INDEX